SOCIAL ROBOTS

Emerging Technologies, Ethics and International Affairs

Series editors:

Steven J. Barela, University of Geneva, Switzerland,
Jai C. Galliott, The University of New South Wales, Australia
Avery Plaw, University of Massachusetts, USA
Katina Michael, University of Wollongong, Australia

This series examines the crucial ethical, legal and public policy questions arising from or exacerbated by the design, development and eventual adoption of new technologies across all related fields, from education and engineering to medicine and military affairs.

The books revolve around two key themes:

- Moral issues in research, engineering and design.
- Ethical, legal and political/policy issues in the use and regulation of technology.

This series encourages submission of cutting-edge research monographs and edited collections with a particular focus on forward-looking ideas concerning innovative or as yet undeveloped technologies. Whilst there is an expectation that authors will be well grounded in philosophy, law or political science, consideration will be given to future-oriented works that cross these disciplinary boundaries. The interdisciplinary nature of the series editorial team offers the best possible examination of works that address the 'ethical, legal and social' implications of emerging technologies.

Forthcoming titles:

Surveillance Technologies
Understanding Their Application to Children and Childhood
Edited by Emmeline Taylor and Tonya Rooney

Drones and Responsibility
Legal, Philosophical and Socio-Technical Perspectives on
Remotely Controlled Weapons
Edited by Ezio Di Nucci and Filippo Santoni de Sio

Social Robots
Boundaries, Potential, Challenges

Edited by

MARCO NØRSKOV
Aarhus University, Denmark

Routledge
Taylor & Francis Group

LONDON AND NEW YORK

First published 2016 by Ashgate Publishing

Published 2016 by Routledge
2 Park Square, Milton Park, Abingdon, Oxon OX14 4RN
605 Third Avenue, New York, NY 10017

First issued in paperback 2021

Routledge is an imprint of the Taylor & Francis Group, an informa business

Publisher's Note
The publisher has gone to great lengths to ensure the quality of this reprint but points out that some imperfections in the original copies may be apparent.

British Library Cataloguing in Publication Data
A catalogue record for this book is available from the British Library

The Library of Congress Cataloging-in-Publication data
Names: Nørskov, Marco, editor of compilation.
Title: Social robots : boundaries, potential, challenges / edited by Marco Nørskov.
Description: Farnham, Surrey, UK ; Burlington, VT : Ashgate, 2016. | Series:
   Emerging technologies, ethics and international affairs | Includes
   bibliographical references and index.
Identifiers: LCCN 2015034826| ISBN 9781472474308 (hardback) |
Subjects: LCSH: Human-robot interaction--Philosophy.
Classification: LCC TJ211 .S654 2016 | DDC 629.8/924019--dc23
LC record available at http://lccn.loc.gov/2015034826

ISBN 13: 978-1-03-209818-0 (pbk)
ISBN 13: 978-1-4724-7430-8 (hbk)

Contents

Notes on Contributors   vii
List of Abbreviations   xi
Acknowledgments   xiii
Editor's Preface   xv

PART I   BOUNDARIES

1   On the Significance of Understanding in Human–Robot Interaction   3
Julia Knifka

2   Making Sense of Empathy with Sociable Robots: A New Look at the "Imaginative Perception of Emotion"   19
Josh Redstone

3   Robots and the Limits of Morality   39
Raffaele Rodogno

4   What's Love Got to Do with It? Robots, Sexuality, and the Arts of Being Human   57
Charles M. Ess

PART II   POTENTIAL

5   Ethics Boards for Research in Robotics and Artificial Intelligence: Is it Too Soon to Act?   83
John P. Sullins

6   Technological Dangers and the Potential of Human–Robot Interaction: A Philosophical Investigation of Fundamental Epistemological Mechanisms of Discrimination   99
Marco Nørskov

7   The Uncanny Valley: A Working Hypothesis   123
Adriano Angelucci, Pierluigi Graziani, Maria Grazia Rossi

8   Staging Lies: Performativity in the Human–Robot Theatre play
       I, Worker   139
       Gunhild Borggreen

PART III   CHALLENGES

9   Robots, Humans, and the Borders of the Social World   157
       Hironori Matsuzaki

10   The Diffuse Intelligent Other: An Ontology of Nonlocalizable
       Robots as Moral and Legal Actors   177
       Matthew E. Gladden

11   Gendered by Design: Gender Codes in Social Robotics   199
       Glenda Shaw-Garlock

12   Persuasive Robotic Technologies and the Freedom of Choice
       and Action   219
       Michele Rapoport

Notes on Contributors

Adriano Angelucci is a Temporary Lecturer in Epistemology at the University of Urbino, Italy. His present research focuses mainly on epistemology and the use of thought experiments in analytic philosophy. He completed his PhD in Philosophy at the University of Verona, Italy, in 2012, after a two-year stay as a visiting researcher at Columbia University, New York. He is currently under contract with an Italian academic publisher (Carocci), for which he is writing an introduction to the use of thought experiments in analytic philosophy. The volume is entitled *Che cos'è un esperimento mentale?* (What is a Thought Experiment?) and is forthcoming in 2016.

Gunhild Borggreen is an Associate Professor in Visual Culture at the Department of Arts and Cultural Studies, University of Copenhagen, and focuses in her research on gender, nationhood, and performance in contemporary Japanese art and visual culture. Borggreen is the co-founder and project manager of ROCA (Robot Culture and Aesthetics; ikk.ku.dk/roca), a research network focusing on practice-based research into the intersection of science, technology, and art. Borggreen has been invited to speak on robots and visual arts in Japan at international conferences, and she has an expanded network of collaborators among international scholars and artists. Borggreen has published in journals such as *Performance Research* and *Copenhagen Journal of Asian Studies*, and is co-editor of and a contributor to *Performing Archives/Archives of Performance* (Museum Tusculanum Press, 2013).

Charles M. Ess is Professor in Media Studies, Department of Media and Communication, University of Oslo. Ess has received awards for excellence in teaching and scholarship; he has also held several guest professorships in Europe and Scandinavia—most recently in the Philosophy Department, University of Vienna (2013-2014). Ess has published extensively in the field of information and computing ethics (for instance *Digital Media Ethics*, 2nd edition, Polity Press, 2013) and in the field of Internet studies (for instance with William Dutton, The Rise of Internet Studies, *New Media and Society* 15(5), 2013). Ess frequently leads workshops on (Internet) research ethics, and is an ethics advisor for international research projects such as the Virtual Centre of Excellence for Research in Violent Online Political Extremism (voxpol.eu) and Existential Terrains: Meaning, Memory, Connectivity (http://et.ims.su.se/). Ess emphasizes cross-cultural approaches to media, communication, and ethics, focusing especially on virtue ethics and the light it sheds on what it means to be human in an (analog-) digital age.

Matthew E. Gladden is interested in fields including social robotics, managerial robotics, and artificial life. His research has been published in *Frontiers in Artificial Intelligence and Applications*, *Business Informatics*, the *International Journal of Contemporary Management*, and the *Annals of Computer Science and Information Systems*, as well as by MIT Press. He is a Research Affiliate at Georgetown University and directs a research and consulting firm specializing in management cybernetics and the development of information security strategies and protocols for neuroprosthetic systems and brain–computer interfaces. He previously served as Associate Director of the Woodstock Theological Center at Georgetown University and taught philosophy at Purdue University. He holds a Bachelor's degree in philosophy from Wabash College and is completing an MBA in data analysis and innovation with the Institute of Computer Science of the Polish Academy of Sciences.

Pierluigi Graziani received his PhD from 'La Sapienza' University of Rome in 2007 and is a Postdoctoral Fellow in Logic and Philosophy of Science at the Department of Philosophical, Pedagogical and Economic-Quantitative Sciences of the University of Chieti-Pescara. His research topics are in the philosophy of computing, mathematical logic, the philosophy of mathematics, and the history of mathematics. The main results of his research have been published in journals and books by Italian and leading international publishers. Graziani is also the co-editor with L. Guzzardi and M. Sangoi of the book *Open Problems in Philosophy of Sciences* (College Publications, 2013), and co-editor with C. Calosi of the book *Mereology and the Sciences: Parts and Wholes in the Contemporary Scientific Context* (Synthese Library, Springer, 2014). Graziani is currently the editor in chief of *Isonomia Epistemologica* and a member of the editorial board of *APhEx*.

Julia Knifka is a PhD candidate as well as a Teaching and Research Associate at the Department of Philosophy at the Karlsruhe Institute of Technology (KIT). She studied at KIT and Yale University and graduated from KIT with an MA degree in European Culture and History of Ideas. In 2014, funded by the Japan Society for the Promotion of Science, she was a visiting researcher at the Department of History and Philosophy of Science at the University of Tokyo. Her research focuses on the social-philosophical and phenomenological implications of technology, more specifically social human–robot interaction. Additional fields of interests include philosophical anthropology, the philosophy of technology and phenomenology.

Hironori Matsuzaki is a sociologist at the University of Oldenburg, Germany. His research interests include social theory, philosophical anthropology, the sociology of scientific knowledge, the sociology of law, and science and technology studies. He has conducted ethnographic research on robotic human science and social robotics in the framework of the "Development of Humanoid and Service Robots: An International Comparative Research Project – Europe and Japan", funded by the DFG German Research Foundation (2010-2014). Since 2014, he has been a member of the Ethics Advisory Board of the EU-funded "Human Brain Project" (Future and Emerging Technologies Flagship). Recently, he has co-edited the special issue *Going beyond the Laboratory: Reconsidering the ELS Implications of Autonomous Robots* (*AI & Society*, 2015, in print).

Marco Nørskov is an Assistant Professor at the Department of Philosophy and History of Ideas at Aarhus University, Denmark. Furthermore, he is a cooperative researcher at the Hiroshi Ishiguro Laboratories, ATR, Japan. He received his PhD in philosophy in 2011 and an MSc in mathematics and philosophy in 2007 from Aarhus University. Nørskov is a member of the PENSOR project (Philosophical Enquiries into Social Robotics) and the administrative coordinator of the TRANSOR network (Transdisciplinary Studies of Social Robotics). His research is focused on robo-philosophy, Japanese thought, phenomenology, and the philosophy of technology. Amongst his recent publications are *Revisiting Ihde's Fourfold "Technological Relationships": Application and Modification* (Philosophy & Technology, 2015), and with Johanna Seibt and Raul Hakli he co-edited *Sociable Robots and the Future of Social Relations: Proceedings of Robo-Philosophy 2014* (IOS Press Ebooks, 2014).

Michele (Michal) Rapoport is a Visiting Graduate Fellow at the J. Edmond Safra Center at Harvard University (2014-15) and was, in 2013-14, a Graduate Fellow at the J. Edmond Safra Center at the Tel Aviv University in Israel. Her PhD in philosophy, awarded in 2015, engages in the philosophy and ethics of intelligent technologies and smart environments; and her research, both in her dissertation and in peer-reviewed publications, focuses on continued visibility and ubiquitous surveillance, the corporeality of the body within digitalized environments, technology and its impact on space and place formation, and the changing nature of agency, labor, free will and personal autonomy in interactions with intelligent devices. Rapoport is a trained architect and incorporates her professional experience to create a unique space for the confluence of questions pertaining to space, ethics, technology, materiality, and an emerging posthuman condition.

Joshua D.B. Redstone is a PhD candidate at Carleton University's Institute of Cognitive Science in Ottawa, Canada. His research focuses on issues that emerge where the fields of social robotics, cognitive science, and the philosophy of mind intersect. He became interested in social robotics and the uncanny valley phenomenon while studying philosophy under Wayne Borody at Nipissing University. He continued to study philosophy at Carleton University, where, under the supervision of Heidi Maibom, he became interested in empathy's role in social robotics. His other research interests include the emotions, the philosophy of technology, experimental philosophy, and consciousness.

Raffaele Rodogno, PhD (St Andrews, Scotland), Associate Professor in Philosophy, Aarhus University, works at the intersection of ethics, the philosophy and psychology of the emotions, and a number of applied issues, including social robotics, criminal punishment, restorative justice, intercultural conflicts, environmental ethics, and euthanasia. He has recently edited a special issue of *Ethics and Information Technology* on ethics and social robots and co-authored *In Defense of Shame* (Oxford University Press, 2011) with J. Deonna and F. Teroni. In the field of social robotics he is particularly interested in the nature of human–robot relations and the effects that the latter may have on human–human relations and practices.

Maria Grazia Rossi is a Postdoctoral Researcher at the Catholic University of Milano (Italy) and an Adjunct Professor at the University of Cagliari (Italy). She obtained her PhD in cognitive sciences from the University of Messina (Italy). Her research interests include the philosophy of psychology, with a focus on language learning and language evolution, moral psychology, and the theory of emotions. More recently she has worked on a theory of emotions in connection with moral and political decision-making. On these issues she has published papers and books, including *Il giudizio del sentiment: Emozioni, giudizi morali, natura umana* (The judgment of feeling: Emotions, moral judgments, human nature; Editori Riuniti University Press). She is currently focusing on the role of reasoning and argumentation in the social and collaborative dimension of human communication (unicatt.it/healthyreasoning).

Glenda Shaw-Garlock is a PhD Candidate at Simon Fraser University (Canada) in the School of Communication within the Faculty of Communication, Art and Technology. Working under the direction of Dr. Richard Smith, Shaw-Garlock's research interests include the cultural and social history of automatons and robots within different cultures; the gendering of technological artifacts; the emergence of sociable and emotive technology in society; as well as critical perspectives on science, technology, and culture. Shaw-Garlock is currently in the final stages of completing her dissertation work and holds Master's and Bachelor's degrees in communication from Simon Fraser University.

John P. Sullins, (PhD, Binghamton University (SUNY), 2002) is a Professor at the Sonoma State University in California where he is a member of the Center for Ethics Law and Society. He teaches philosophy and robotics, cyberethics, philosophy of science and technology and logic. His research interests are in the philosophy of technology, computer ethics and the philosophical implications of technologies such as robotics, AI and artificial life. His recent publications have focused on artificial agents and their impact on society as well as the ethical design of successful autonomous information technologies including the ethics of the use of personal robots, autonomous vehicles, and robotic weapons systems.

List of Abbreviations

A-life	Artificial life
AI	Artificial intelligence
AMA	Artificial moral agent
BA	Bodymind awareness
CAP	Computing and Philosophy
GOFAI	Good old-fashioned AI
HRI	Human–robot interaction
IRB	Institutional review board
ICT	Information and computing technologies
IT	Interaction theory
LARs	Lethal autonomous robots
pHHI	Physical human–human interaction
pHRI	Physical human–robot interaction
SIAs	Socially intelligent agents
TEAR	The ethics application repository
THRI	Theater-based human–robot interaction
WoZ	Wizard-of-Oz

Acknowledgments

I would like to thank all the contributors for their hard work, comments, support, and discipline during the editorial phase of this project. Without them the book would simply not have been possible. The initiative for this book is one of the many outcomes of the Robo-philosophy conference, which was held in 2014 at Aarhus University in Denmark. One of the first major conferences of its kind, it attracted about 120 participants including many of the pioneers within the field. I would like to use this occasion to thank my co-organizers Johanna Seibt and Raul Hakli, as this conference was a significant first step towards this book, as reflected in the fact that several of the contributions of this volume to various extend are based on earlier versions of the conference proceedings (2014). Furthermore, I would especially like to thank the co-editor of Emerging Technologies, Ethics and International Affairs Series, Jai Galliot, as well as the publisher of Ashgate's Politics and International Relations Kirstin Howgate and her assistant editor Brenda Sharp for their encouragements, guidance, flexibility, and patience. I am greatly obliged to my dear colleagues Johanna Seibt and Raffaele Rodogno from Aarhus University, who were always there when I needed advice or sparring partners to discuss various editorial questions, as well as to Maja Kragh-Schwarz from the Technology Transfer Office at Aarhus University for her thorough legal advice. Last but not least, I would like to thank my wife Sladjana and my son Leon for their tolerance on all those many occasions when work had the highest priority in order to meet the various deadlines in the publication process.

This volume was made possible by the VELUX Foundation's support of the PENSOR research project at Aarhus University.

Editor's Preface

By exploring the potential of social robotics we are currently stretching the very plasticity and boundaries of our social interaction space, which in our contemporary age is often conceived of as reserved for humans and their relations with each other. The advent of robots as novel interaction "partners", whether as supplements or substitutes, challenges our self-understanding as individuals and as collectives on various levels. Recent anecdotal indications of this tendency are represented by the so-called *robotic moment*, a term coined by Sherry Turkle.

> I find people willing to seriously consider robots not only as pets but as potential friends, confidants, and even romantic partners. We don't seem to care what these artificial intelligences 'know' or 'understand' of the human moments we might 'share' with them. At the robotic moment, the performance of connection seems connection enough. (Turkle 2012, p. 9)

And there is more to it, as the qualitative interaction experience with social machines might not only be enough but may even be preferred to its human–human interaction equivalents (cf. ibid.). It is in the context of these recent technological developments and its prospects that this volume aims to explore the boundaries, potential, and challenges pertaining to the meeting with novel types of social otherness.

The contributions to the anthology at hand comprise an interwoven and cohesive inquiry into the field of social robotics. Its main vantage point is philosophical, however, as social robotics itself is a truly interdisciplinary undertaking, the chapters also bring into play diverse strains of thoughts from different disciplines such as sociology and aesthetics. Needless to say, such an inquiry can hardly be exhaustive. As noted in Chapter 6 the term "robot" is rather vague and used for a variety of machines, which differ considerably in their nature—for instance vacuum-cleaning robots, teleoperated drones, science fiction robots. "Social robotics" is no exception and the term does not only apply to humanoid robots. While most of the contributions in this volume take robots with great human affinity as the main focus of their inquiry, several of the chapters also address nonhuman shaped applications and interaction contexts such as pet robots, augmented social robotic technologies, and nonlocalizable robotic systems. All these contributions aim at providing insight and a thorough understanding of selected facets of human–robot interaction (henceforth: HRI). An in-depth understanding of the issues addressed here is crucial if the research and development of social robotics is to result in sustainable and responsible robotic applications in social contexts.

Due to the potentially disruptive nature of these technologies, it is natural that the majority of philosophical literature focuses on ethics. Although several of the chapters also deal with ethical and moral issues, the scope of the topics and questions raised here is broader. The label "robo-philosophy", as defined, probably for the first time, by Johanna Seibt (cf. 2014), is more appropriate, as the contributions bring into play a wide range of philosophically related strains of thought with the ambition of opening up future discussions that feed back into theory as well as praxis and go beyond applying existing philosophical theory to robotic phenomena.

The nascence and future orientation of the field imposes some challenges on any inquiry. By constantly pushing the limits of what is technically feasible and conceivable, many of the short- and long-term applications as well as their contexts are conceptual by nature. This does not make them less relevant. On the contrary, it is vital to contemplate the boundaries, potential, and challenges of potentially highly disruptive technologies in order to exercise due diligence and to ensure a prosperous development of all stakeholders, ranging from the individual to businesses and organizations, to societies as a whole, and to human kind.

Among other things, the topics addressed in this volume concern socio-political issues, social perception, cognition, and recognition, and the questions of how to conduct and progress with research and development in the field of social robotics. The chapters are interrelated in multiple ways; nevertheless, I hope that the reader will find the segmentation of the volume and the overview below helpful.[1]

Part I: Boundaries

Given the current state of the art within the field of robotics and its short-term prospects, Part I explores the boundaries of HRI. More concretely, it investigates central issues on concrete and conceptual levels such as cognitive capacity requirements for genuine social interaction and the phenomenon of empathy with social robots; as well as discussing whether or not robots can be moral agents, and exploring the limits of robotics when it comes to intimate relations.

Certain minimal requirements seem to be prerequisite if robots are to enter fully into our social sphere *on a par* with other social actors. In Chapter 1, "On the Significance of Understanding in Human–Robot Interaction", Julia Knifka explores the cognitive capacities necessary for such a robot. Drawing on Alfred Schütz's socio-phenomenology, she examines the importance of robots' understanding of observable body movements and actions of their human interaction partners. More concretely, she argues that it would not be sufficient for a robot only to observe a human's behaviors and to be able to mimic social cues. For a robot to become

---

1   References in the overview of the parts and chapters are only added to those passages that bridge them together, as the relevant citations pertaining to the contents of the respective chapters are to be found there.

an intelligible social companion, it would also need the capacity to understand and interpret all this information as well as the relevant situational contexts—in addition to reacting adequately as a result. She concludes that genuine social interaction with a robot will not be possible unless the robot is able to comprehend our perspective in order to engage in joint actions that are based on a mutual understanding of the other person's underlying motives.

In Chapter 2, "Making Sense of Empathy with Sociable Robots", Josh Redstone examines Catrin Misselhorn's notion of empathy towards inanimate objects, which is a product of the interplay between our imagination and perception. Redstone points to the semantic ambiguity of the concept of "imagination", and argues that empathy here would be more adequately understood in terms of a perceptual illusion. This is more a constructive refinement of Misselhorn's theory than a refutation of it. Based on his finding, Redstone modifies Misselhorn's framework and tests it on Mori's *Uncanny Valley* hypothesis, as Misselhorn links the feeling of eeriness with the failure of empathy. He suggests that in the modified framework, the eeriness that we might experience in the encounter with very humanlike robots turns out to be based not so much on a failure but on the persistence of empathy. The chapter concludes with a discussion on the status of empathy in Misselhorn's and in the modified framework. As robots are not in possession of human emotions, for instance, illusory and imaginative empathy do not qualify as genuine empathy.

If robots enter the boundaries of our social domains in roles that are only viable for humans, it becomes imperative to clarify their moral standing in these emerging relations. In Chapter 3, "Robots and the Limits of Morality", Raffaele Rodogno critically discusses the possibility of social robots qualifying as moral agents and patients. Despite arguments by David Gunkel and Mark Coeckelbergh to the contrary, Rodogno concludes that robots are not appropriate objects of moral concern (at least not until they possess certain properties). He argues that in the absence of humanly recognizable interests, it is unintelligible for human beings to know what could ground moral consideration. Rodogno, however, does not want to exclude the possibility that robots may be the bearers of such interests. As for moral agency, Rodogno adopts a form of moral sentimentalism according to which the capacity to experience certain emotions is necessary to gain central moral insights. For as long as robots are not capable of such emotions, they will not be full-blown moral agents. In short, in Rodogno's view, both moral agency and patiency are concepts whose structure and scope are limited by human biology and its cultural declinations.

Our intimate spheres are not sacrosanct domains when it comes to the application contexts and marketplaces targeted by social robotics. In Chapter 4, "What's Love Got to Do with It?", Charles M. Ess explores the very limits of HRI by inquiring into the status of sexual engagement with robots. He argues that although certain sexual praxes with robots seem conceivable, robots will nonetheless remain incapable of truly engaging with us as romantic partners with whom we can realize what Sara Ruddick refers to as "complete sex." Lacking full autonomy, genuine emotions and other qualities, robots are insufficient when it comes realizing the adequate level of intimacy, as this is dependent on the mutuality of these innate human features—most specifically, the mutuality of desire as an emotion. In other words, complete sex is marked by a mutually shared desire to be desired—but social robots are incapable of desiring, so no such mutuality of desire is possible. Furthermore, Ess points at the disruptive potential associated with settling for less. The reason is that *complete sex* further realizes and nurtures various virtues such as love and respect. While robots as sex partners might be perfectly legitimate in some contexts, Ess also argues that in the absence of *complete sex*, robots will not inspire and foster these virtues—virtues that are necessary for both deep friendships and *complete sex*. Conflating sex with robots with the potential we can realize in human–human intimacy fosters the danger of depriving us from living a virtuous and good life as both friends and lovers.

Part II: Potential

In spite of its boundaries and challenges, social robotics fosters great possibilities ranging from contributions to the interdisciplinary research field itself to the quality of the life of the end-user, the "H" in HRI. In Part II a variety of these potentials and their prerequisites will be presented. On a practical level, an instrument for ensuring that social robotics develops its potential in a sustainable and responsible way will be introduced. The possibility of self-realization empowered by HRI will be discussed, and it will be suggested that an inquiry into the process of dehumanization might shed new light on the problem of the uncanny phenomenon, which presents a severe limitation when it comes to user experience. Finally, it will also be demonstrated that robotic performances in combination with aesthetics expose central aspects of human nature.

Robotic technologies are already deployed in mission-critical contexts ranging from warfare to healthcare. However, as will be argued in Chapter 12 (Part III), less high-stake, everyday robotic applications also affect us profoundly. Based on short- and long-term technological prospects, John P. Sullins discusses in Chapter 5, "Ethics Boards for Research in Robotics and Artificial Intelligence", the necessity of implementing functional ethical boards in order to effectively safeguard responsible and sustainable research and development when it comes to these disruptive technologies. Contrary to the claim that the current status quo does not warrant such regulatory instruments, Sullins argues that it is actually a rather urgent matter. He constructively outlines the advantages and pitfalls of ethical reviews on an individual and organizational level. A central requirement for Sullins, in order to make an ethical board operational and efficient, is to provide researchers, programmers, and engineers with an adequate level of ethical education in order to ensure the symbiotic flourishing of the individual, society, and industry as a common goal.

The question of how we ascribe capacities such as feelings, beliefs, and desires is closely linked to the way we conceptualize and recognize social robots. The fundamental issue of conceptualization will be addressed in Chapter 6 in "Technological Dangers and the Potential of Human–Robot Interaction." Here Marco Nørskov applies David Shaner's three levels of the Buddhist *bodymind awareness* to the case of HRI. He argues that the underlying epistemological mechanics behind conceptualization with respect to robots is an artifact itself, and highlights the moral significance of this process. The Buddhist-inspired thought applied will be informed by and contrasted with Martin Heidegger's philosophy of technology in order to identify the technological dangers as well as the potential of social robotics. Nørskov furthermore examines the possibility of a mode of interaction with robots which goes beyond a reduction to a mere *standing reserve*, and discusses the possibility of praxes that could foster human self-realization empowered by HRI.

The Uncanny Valley hypothesis, which is also addressed in the previous chapter, will be the main topic of Chapter 7. Masahiro Mori originally published this hypothesis in 1970. Here he recommended that roboticists should refrain from designing too humanlike robots, as there is a risk of producing machines that will induce a feeling of uncanniness or eeriness in the end-user. In the contribution bearing the same title as Mori's paper, *The Uncanny Valley*, Adriano Angelucci, Pierluigi Graziani and Maria Grazia Rossi critically examine how robotics and computer graphics have tried to technologically overcome this phenomenon as it imposes serious challenges with respect to user experience. Lacking a sufficient level of realism, robots and graphical animations which fall into the uncanny valley are inept when it comes to invoking empathy in their human interaction partners. This poses a significant limitation if these robots are to function as social actors in HRI. Whereas previous attempts mainly focused on the cause of the uncanny feeling, the authors suggest that there is another domain of inquiry that has so far not been thoroughly investigated: the nature of the very feeling of uncanniness itself in the HRI context. The authors emphasize that this is a philosophical endeavor. They further suggest that any reluctance to grant the status of being human to characters falling into the uncanny valley relates to the problem of recognition. It is in this context that they conclude by formulating a new hypothesis, namely that the emotion of disgust is linked to a process of dehumanization. They anticipate that a close examination of the process of dehumanization could provide a more comprehensive understanding of the uncanny valley phenomenon.

Due to the complexity of human social interaction (cf. Part I), teleoperated robots are often used instead of their autonomous counterparts in order to compensate for technological shortcomings (cf. Ishiguro and Nishio 2007, p. 135). Animated in this way, these robots are used in creative ways within HRI research and other areas of application. One such other context is art, where robots for instance perform as 'actors' in theater plays. Gunhild Borggreen demonstrates in Chapter 8, "Staging Lies", how robotic art and aesthetics can feed back into and inform robotics. Her analysis is mainly based on the case of *I, Worker*, a theater production directed by Oriza Hirata in collaboration with Hiroshi Ishiguro featuring two humanoid robots. But she also discusses the concept of the Wizard of Oz as a theater-like HRI test setting in a laboratory. Focusing on lies, a literal lie by a robot in the former case, and a lie which leaves test subjects naïve to the internal mechanisms of the machines in the latter, Borggreen demonstrates how aesthetics reveals fundamental issues regarding what it means to be a human being. Based on John L. Austin's theory, she argues that robots to which we ascribe human capacities act on the human interaction paradigm as 'parasites', as their pretense undermines our established inter-human praxes. In her final analysis, Borggreen concludes that theories of performativity and aesthetics can be an effective means to illustrate the degree to which we are socially 'programed' to respond to the world within certain norms and conventions, and how these can be used to establish agency.

Part III: Challenges

When it comes to the integration of social robotic systems into our society, the ambiguity of the ontological, social, and ethical status of these machines imposes various challenges on our establishments and self-understanding, as well as on our way of living at a fundamental level. Part III examines several of these challenges. More specifically, it will examine the question of how we can empirically analyze interactions with robots when their social status is not settled. Ethical and legal problems concerning nonlocalizable robotic systems will be discussed, as well as issues pertaining to the gendering of robots. Furthermore, the limitations imposed on our volitional and action space by robotic systems and their influence on our autonomy and subjectivity will be disclosed.

Social robotics questions the foundations of our institutional and functional notions of what it means to be social and ultimately a human being—empirically as well as conceptually. In Chapter 9, "Robots, Humans, and the Borders of the Social World", Hironori Matsuzaki examines the challenge that social robotics imposes on our understanding of human–robot-interaction when it comes to sociality, and more specifically on how to conduct critical (field) research. Based on the observation that the notion of "human being" is not culturally or historically stable, as well on the argument that being human is not identical with being social, he suggests that social aspects of HRI should be analyzed in the same way as human–human interaction. As a consequence, and grounded on basic assumptions within sociology, Matsuzaki develops a framework for studying HRI which aspires to suspend from possible anthropological biases. The triadic framework proposed here offers a conceptual instrument which can be used to empirically analyze interactions with robots as well as their agency in borderline cases, where for instance the status of the robot involved is ambiguous to its interaction partner.

Who are we to blame for actions performed by robots—the manufacturers, the owners, the robots or somebody else? As long as we have someone or something to point at, our legal and ethical framework has a concrete vantage point from which to approach this question. Nevertheless, complexity increases when it comes to robots lacking an identifiable physical body. In Chapter 10, "The Diffuse Intelligent Other", Matthew E. Gladden takes up this challenge and develops an ontology of nonlocalizable robots as moral and legal actors based on the their *autonomy*, *volitionality*, and *localizability*. Grounded in this novel ontology, he discusses legal and moral responsibility with respect to robots' actions in traditional contexts, such as tool use and swarm networks, as well as issues that have so far not been rigorously addressed pertaining to nonlocalizable robots. *Robots as Ambient Magic*, *Robots as Diffuse Animal Other*, and *Robots as Charismatic Lawgiver and Moral Beacon* are just some of the types of nonlocalizable robot technologies Gladden considers in his contribution. Finally, he discusses how established ethical and legal theories could be applied to accommodate these new entities.

Sociable robots are deliberately designed to trigger our natural disposition to anthropomorphize objects around us, for instance with respect to the ascription of the capacity to feel. Gender is no exception, and is also one of the cues in the repertoire of the robot engineers that can be used to influence the affection of humans emerging in the interaction with their robotic creations. In Chapter 11, "Gendered by Design", Glenda Shaw-Garlock provides a detailed overview of the current state of affairs on gender research in the context of robotic design based on HRI research, communications studies literature, and science and technology studies. She discusses what it means for a social robot to be gendered, and explores some of the implications associated with the ascription of gender. Various studies demonstrating the effects of gendering robots are examined here, and ethical dimensions are also discussed. Furthermore, several constructive strategies for mitigating stereotyping are presented. As a final point, the importance of robot designers including gender-sensitive approaches in their robot design is stressed.

As pointed out by Martin Heidegger (2000), we face a great risk when we regard technology as something neutral—it forms our very understanding of the world and ourselves. In Chapter 12, "Persuasive Robotic Technologies and the Freedom of Choice and Action", Michele Rapoport demonstrates how even robotic technologies which engage in routine and mundane activities, and which are currently entering domestic markets, limit our volitional and action space and ultimately affect our autonomy and subjectivity. Rapoport argues that high-tech devices such as smart refrigerators, semiautonomous cars, and health-monitoring gadgets influence and control our self-regulating mechanisms. They challenge our personal freedom through nudging, determining, and limiting the domain of possibilities with respect to such central human capacities as decision-making and the consequential realization of actions. Although these devices might appear rather unproblematic and beneficial at first, we risk paying a price by outsourcing these capacities, as the devices' programmed regulatory functionality does not necessarily correspond to our real momentary preferences. Furthermore, pervasive and intelligent technologies are not only supposed to guide us—they are often expected to take over when we do not make the "right" choices, and hence have profound effects on our autonomy and subjectivity. While these new technologies engage in "nudge" and in the disciplining of the subject in what is perhaps a more efficient manner, they also create a new ethical landscape due to their ubiquitousness and intimate engagement in human daily activities.

Marco Nørskov
Aarhus (Denmark), 2015

References

Heidegger, M. 2000. "Die Frage nach der Technik." In *Vorträge und Aufsätze*, edited by Friedrich-Wilhelm von Herrmann, 5-36. Frankfurt am Main: Vittorio Klostermann. Original edition, 1953.

Ishiguro, H. and S. Nishio. 2007. "Building Artificial Humans to Understand Humans." *The Japanese Society for Artificial Organs* 10(3): 133-42. doi: http://dx.doi.org/10.1007/s10047-007-0381-4.

Seibt, J. 2014. "Introduction." In *Sociable Robots and the Future of Social Relations: Proceedings of Robo-Philosophy 2014*, edited by Johanna Seibt, Raul Hakli and Marco Nørskov, vii-viii. Amsterdam: IOS Press Ebooks.

Seibt, J., R. Hakli, and M. Nørskov, eds. 2014. *Sociable Robots and the Future of Social Relations: Proceedings of Robo-Philosophy 2014*, edited by J. Breuker, N. Guarino, J.N. Kok, J. Liu, R. López de Mántaras, R. Mizoguchi, M. Musen, S.K. Pal and N. Zhong, *Frontiers in Artificial Intelligence and Applications*. Amsterdam: IOS Press Ebooks.

Turkle, S. 2012. *Alone Together: Why We Expect More from Technology and Less from Each Other*. New York: Basic Books. Original edition, 2011.

PART I
Boundaries

Chapter 1

On the Significance of Understanding in Human–Robot Interaction

Julia Knifka

This chapter argues that in order to talk about human–robot interaction it is paramount to analyze how and whether humans and robots can understand each other. Deriving from different types of interactions and roles that are ascribed to the robot, this contribution will present different parameters and levels of understanding in regard to human perceptual experiences. It will be shown that understanding in a social interaction is guided by observations made and interpretations of a given situation as well as influenced by personal, cultural experiences of the observer. To this extent, the socio-phenomenological approach of Alfred Schütz on human–human interaction will shed light on the possibilities and problems of understanding in human–robot interaction.

Introduction

We assume that human interactions with robots in the future will be different from the way we use simple technological artifacts today due to technological advances in imitating or reproducing human behavior. While interaction originally referred to multi-agent systems, the use of the term shifted from a technical description of what happens within a system towards a social description of the relation between humans and robots. Talking about human–robot interaction is common practice in robotics. A very special form of HRI is social robotics focusing on social interaction. The idea of a social robot that assumes the role of a social partner or companion is associated with the belief that a robot will be generally accepted when it appears like a human (Kanda et al. 2007, Fong, Nourbakhsh, and Dautenhahn 2003) and "displays rich social behavior" (Breazeal 2004, p. 182). The long-term aim is to build "human-made autonomous entities that interact with humans in a humanlike way" (Zhao 2006, p. 405). Looking and behaving like a human is to create the impression that "[i]nteracting with the robot is like interacting with another socially responsive creature that cooperates with us as a partner" (Breazeal 2004, p. 182). Jutta Weber (2014)—in analogy to Searle's distinction in the field of AI (see Searle 1980)—distinguishes between a strong and a weak approach of HRI. The weak approach in HRI follows the assumption that emotions, sociality, and aspects of being human that finally would require a theory of mind cannot be realized in robotics, but can only be imitated and made believe. The strong HRI approach, by contrast, is based on man's ability to create robots that are social in their nature:

> The strong approach in HRI aims to construct self-learning machines that can evolve, that can be educated and will develop real emotions and social behavior. Similar to humans, social robots are supposed to learn via the interaction with their environment, to make their own experiences and decisions, to develop their own categories, social behaviors and even purposes. ... [T]he followers of the strong approach—such as Cynthia Breazeal and Rodney Brooks ... —strive for true social robots, which do not fake but embody sociality. (Weber 2014, p. 192)

Apart from the possibility of robots emerging as a complete different species with their "own categories", "embody sociality" (ibid.) indicates that robots will become social beings and, moreover, develop a mind equal to the human mind through emerging cognitive behavior and interaction. An important aspect implied is that the robot will be able to understand us. Without understanding, no interaction is possible, since social interaction between humans is a multifaceted phenomenon with various layers of verbal and non-verbal behavior that need to be understood and interpreted in the context of the situation. The significance of the observation of bodily movements for understanding a surface behavior at least will be discussed in this contribution.

The notion of understanding has a broad history and methodology in philosophy, especially hermeneutics, which is the philosophy of understanding. Alfred Schütz, whose concepts may be useful for analyzing the question of understanding between humans and artifacts, was an intellectual disciple of Husserl's phenomenology. He wanted to put Weber's interpretative sociology on a phenomenological foundation, but did not consider his work to be rooted in hermeneutics.[1] Nevertheless, he avails himself of fundamental hermeneutic concepts (cf. Staudigl 2014, p. 3). Understanding is one of them. The process of understanding proceeds along the interaction with the alter ego and is not reduced to mere subjective acts of understanding. Two aspects make Schütz's approach suitable for analyzing whether there can be an understanding between humans and robots at all: His conception of the construction of our social reality is pragmatic in nature and, hence, constituted of social action and interaction. And additionally, the interaction partner does not necessarily have to be—in the strict sense—a conscious being. Schütz comprehends the social world to be constituted of interaction processes in which the subject develops his thought, perception, and actions in correlation with the counterpart and reconciliation of existing sociocultural patterns. He focuses on the formation of meaningful action and asks how the formation processes of understanding of action and conduct emerge in

---

1   He never mentioned hermeneutics and due to his premature death, was probably not familiar with the emerging field of "hermeneutical phenomenology."

daily life and how mutual understanding and relations develop in general (Grathoff 1989, p. 181).

Initially, I will give an overview of different approaches to and visions of human–robot interaction so as to characterize the role understanding plays in HRI. It is claimed that understanding should go beyond the mere observation of behavior. However, observations will be shown to proceed along different levels of understanding. During an interaction, the observation of behavior is always accompanied by anticipations of future actions and an interpretation of the situation at hand. This will be demonstrated with concepts based on the socio-phenomenological concepts of Alfred Schütz.

Interaction with Robots

In his studies of human–robot interaction while dancing a waltz, Kosuge distinguishes between high-level and low-level interactions:

> The goal of our research is to reproduce the pHHI of waltz with pHRI, in which a mobile robot plays the follower's role, being capable of estimating human leader's next step (higher level) and adapting itself to the coupled body dynamics (lower level). (Wang and Kosuge 2012, p. 3134)

Couple dances are inherently social, with the waltz serving as a "typical example of demonstrating human's capabilities in physical human–human interaction (pHHI)" (ibid.). Here, the low-level physical Human–Robot Interaction (pHRI) is viewed from a Newtonian perspective and considered to be a contact force, a force applied to each other. High-level interaction, which in Wang and Kosuge (2012) is second-tier, still needs the robot to have advanced motor skills or to "know" how to a) move by itself and b) estimate/anticipate the next move (ibid., see also Ikemoto, Minato, and Ishiguro 2008, p. 68). However, Kosuge acknowledges that:

> [H]owever, since most of these robots have been developed for entertainment or mental healing, we could not utilize them for realizing complicated tasks based on the physical interaction between the humans and the robots. To realize complicated tasks, however, physical interaction between the robots and the humans would be required. (Kosuge and Hirata 2004, p. 10)

> In the case of coordination among humans, each human would move based on the intention of other people, information from environments, the knowledge of executed tasks, etc. If the robots could move based on the information actively similar to the humans, we could execute various tasks effectively based on the physical interaction with the robots. (ibid.)

Kosuge and Hirata (2004) emphasize that physical interaction based on bodily movements is not only the execution of physical forces, but highly influenced by cognitive capacities, such as intentions, information gathered from the direct environment, and also the knowledge of how to apply or classify the information.

An even more basic approach is chosen by Ikemoto, Minato, and Ishiguro (2008), who go a step back and develop "a control system that allows physical interaction with a human, even before any motor learning has started" (ibid., p. 68): The success of the interaction, which consists of a human helping the robot CB^2 to stand up, is measured by three categories of smoothness (smooth, non-smooth, failed; ibid., p. 69). Despite the basal, prerequisite approach to interaction, pHRI is an "extension to HRI" and in the long run "this research is to allow a humanoid robot to develop both motor skills and cognitive abilities in close physical interaction with a human teacher" (ibid., p. 72).

The reverse objective, which amplifies the close, even more dynamic interaction with a human teacher, is to deploy a robot as a teacher itself.

At this point, another area of HRI research is entered. It focuses on assigning specific roles to the robot. Scholtz (2003) created a taxonomy of roles the robot can assume in a work environment: Supervisor, operator, mechanic, peer, and bystander. All roles are defined by the amount of control exercised over the robot. Scholtz, unlike many others in HRI research, explains how he defines human–robot interaction:

> I use the term "human–robot interaction" to refer to the overall research area of teams of humans and robots, including intervention on the part of the human or the robot. I use "interventions" to classify instances when the expected actions of the robot are not appropriate given the current situation and the user either revamps a plan; gives guidance about executing the plan; or gives more specific commands to the robot to modify behavior. (Scholtz 2003, sec. 2)

The quote demonstrates that, in this case, human–robot interaction is asymmetric and not designed to be social, but task- and goal-oriented in order to achieve joint results. Goodrich and Schultz (2007, pp. 233-4) extend the scope of the taxonomy by the roles of information consumer ("the human does not control the robot, but the human uses information coming from the robot") and mentor ("the robot is in a teaching or leadership role for the human"). The roles still are asymmetric, which is due to the specific nature of these roles. Nevertheless, the robot is given a higher function by reversing the roles.[2] Especially the roles of peer and mentor are appealing, because the robots can act as assistant robots in a wide range of implementations. Examples are Robovie in a classroom (Kanda et al. 2007, Chin, Wu, and Hong 2011), Robotinho—a mobile full-body humanoid museum guide (Faber et al. 2009, Nieuwenhuisen and Behnke 2013), or KASPAR, the social

---

2   Even though the human still is the last instance to control the robot, the human is subordinate when the robot is mentor or teacher. Role conflicts can beexpected.

mediator for children with autism (Dautenhahn 2007, Iacono et al. 2011). By assigning the roles of a social mediator, mentor, or peer to a robot, the emphasis is placed on the social aspects of the interaction. It becomes obvious here that the question of what constitutes social does not seem to be relevant: Social interactions only seem to refer to the fact that the situation the human finds itself in is social in regard to an occurring encounter with another entity, in this case, a robot. The robot notices the human and can answer questions relating to its role. A social conduct, such as greetings and, in general, a social behavior seems to be an addition. Guiliani et al. (2013) even asked whether social behavior is necessary for a robot. To answer this question, they conducted an experiment to compare task-based behavior with socially intelligent behavior of a robotic bartender.[3] The interesting aspect is that bartending is a service, but can also play a social role. Especially when it comes to role ascriptions, developing role-appropriate personas has recently become more relevant. Due to necessity—because collecting data on individual profiles was found to be difficult—personas represent a potential group of subjects. In the long run, the idea is that the robot already displays certain social behavior and can then adapt to the individual needs of its owners. Sekman and Challa, for example, outline:

> A social robot needs to be able to learn the preferences and behaviors of the people with whom it interacts so that it can adapt its behaviors for more efficient and friendly interaction. (2013, p. 49)

However, Dautenhahn emphasizes:

> However, it is still not generally accepted that a robot's social skills are more than a necessary 'add-on' to human–robot interfaces in order to make the robot more 'attractive' to people interacting with it, but form an important part of a robot's cognitive skills and the degree to which it exhibits intelligence. (2007, p. 682)

Increasing the robot's attractiveness is supposed to be achieved by making the robot socially more intelligent without the human tendency of anthropomorphization (Dautenhahn 1998, p. 574). For this purpose, the "socially intelligent hypothesis", which emanates from the idea that intelligence originally had a social function and later was used to solve abstract (e.g. mathematical) problems, should be applied to robotics (Dautenhahn 1998, 1995, 2007). In fact, Dautenhahn states that

> ... socially intelligent agents (SIAs) are understood to be agents that do not only from an observer point of view behave socially but that are able to recognize and

---

3   This research shows that social perception highly depends on gender, age, and nationality.

identify other agents and establish and maintain relationships to other agents. (Dautenhahn 1998, p. 573)

Embodying social behavior goes beyond verbal interactions or the display of emotions, but includes developing an own personal narrative for an individualized interaction ("autobiographic agent" ibid., p. 585) and to enhance social understanding. Social understanding is not only grounded in a biographic context, it also needs "empathic resonance" (ibid.). These central aspects of social understanding parallel the paradigms for a safe social interaction between humans and robots introduced by Breazeal, namely, readability, believability, and understandability (Breazeal 2002, pp. 8-11).

The notion of "readability" is to guarantee that the robot provides social cues for the human to predict the behavior and the actions of the robot. This means that its modes of expression, such as facial expressions, eye gaze, etc., must be revealing and understood easily, so that

> … the robot's outwardly observable behavior must serve as an accurate window
> to its underlying computational processes, and these in turn must be well
> matched to the person's social interpretations and expectations. (ibid., p. 10)

Thus, the computational processes must mirror the behavior in process. The underlying assumption is that humans are readable and that readability is an indicator of the social behavior. The underlying processes of human behavior, namely, human thinking acts are not readable for the human eye and moreover, readability might not be true for all social interactions. Most interactions rely on very subtle social cues, which are happening within seconds, e.g. micro-expressions. Those might not even be visible or consciously noticeable, but humans notice them subconsciously and they might determine our future interaction with the other human partner. Readability is important for the human to react to the robot's behavior. The behavior must create the "illusion of life" (ibid., p. 8). Breazeal calls this believability. The "illusion of life" is made believable, not only because the robot appears to be alive, but also because it displays personality traits (ibid., p. 8, see also Dautenhahn 1995, 1998). The believability highly depends on the user: To be believable, an observer must be able and willing to apply sophisticated social-cognitive abilities to predict, understand, and explain the character's observable behavior and inferred mental states in familiar social terms (Breazeal 2002, p. 8). Although social interactions always depend on the willingness of the participating parties, the robot must display a certain characteristic human-like behavior to increase the willingness. To interact with people in a human-like manner, sociable robots must perceive and understand the richness and complexity of natural human social behavior. Humans communicate with each other through gaze direction, facial expression, body movement, speech, and language, to name a few (ibid.). Hence, the robot must not only display believable social cues, but also understand them and react to them. Breazeal calls this feature "human awareness" (ibid., p. 9). Human awareness is not only restricted to the aspects of communication (including emotions), but includes the array of situational, personal, cultural, and historical contexts as well. To emulate human social perception, a robot must be able to identify who the person is (identification), what the person is doing (recognition), and how the person is doing it (emotive expression). Such information could be used by the robot to treat the person as an individual, to understand the person's surface behavior, and to potentially infer something about the person's internal states (e.g., the intent or the emotive state) (ibid.). The robot's behavior includes communication, interaction skills, and the ability to relate to a human being. Breazeal extends her idea as follows:

> Such a robot must be able to adapt and learn throughout its lifetime, incorporating shared experiences with other individuals into its understanding of self, of others, and of the relationships they share. (ibid., p. 1)

In order to interact with a human in an intuitive and natural manner, a robot must not only understand the human agent. Breazeal also points out that such a robot must be able to understand itself. Understanding oneself in this case means to understand one's "own internal states in social terms" (ibid., p. 19). Besides understanding the intentions, beliefs, and desires of human beings, it also must understand its own intentions, beliefs, and desires.

So far, we have seen that interaction is considered from different perspectives: From a mere physical point of view to the ascription of roles and personas to social understanding through artificial social intelligence. In the first instance movements are calculated in terms of physical force as well as step sequence, cognitive components are excluded along with phenomenological aspects pertaining to bodily movements and orientation within the world. Although a dance like a waltz, which follows a determined step sequence, requires an appropriate proxemics distance and is regarded to be rather formal, it is still a necessity that the involved bodies adapt themselves to each other. The two participating parties have to find a common rhythm, adjust, and synchronize their bodily movements (see also Janich 2012).

The interaction between a human and a robot that embodies a specific role and might even be equipped with an appropriate persona focuses more on the specific social setting. The expectations on the robot are guided and determined by the roles. Personas reproduce adequate behaviors for the roles. It is different with robots in the role of social companions that are supposed to meet on level terms: Here, the development of social intelligence is required or must elicit intuitive and natural behavior and the robot must be able to understand humans and itself to a certain degree.

Levels of Understanding: Observing Bodily Movements and Actions

Regarding interaction, the role of embodiment (cf. Dreyfus 1972) and the common idea of "embodiment interaction" (Dourish 2001, Gallagher 2013) pervade research along with the notion that cognitive capacities are the consequence of a combination of corporeal capacities embedded in a social setting and will eventually emerge—given the right combination. Analogously, Gallagher argues that a "robot will already have to be an IT [interaction theory, JK] robot, if it is required to function in the seamless and relatively reliable way that characterizes most human interactions" (Gallagher 2013, p. 457). Elsewhere he states that a robot capable of smooth embodied interaction with a human solves the problem of social cognition without any theory of mind (ibid., p. 458). Leaving the implications of this aside, I agree with Gallagher that interaction depends on "behavior that is recognizable and understandable in an embodied (primarily inter-subjective) way" (ibid., p. 487). By reference to developmental studies, Gallagher emphasizes that humans are not detached observers, but early on find themselves

> ... drawn into second-person interaction in the embodied practices of primary intersubjectivity. ... Even in instances when we are offline observers, our embodied experiences with others, which give us easy access to what has been called the "massive hermeneutical background" provided by those early and continuing experiences with others, provides the starting point for our understanding. (ibid.)

Gallagher mainly confirms phenomenological insights. When taking a closer look into the phenomenological tradition with a specific focus on a second-person observation and peripheral inter-subjectivity, it becomes obvious what "massive hermeneutical background" (ibid., also Gallagher 2001) means.

If it comes to social interaction, we have to consider that interaction partners are—when it comes to humans—subjects who are socially related to each other. Strictly speaking, interaction is an inter-subjective execution of a joint action. Inter-subjective means a common as well as a shared connection to and engagement in the world. It requires an understanding of the shared social and cultural practices as well as an understanding of the other subject. This shared sociocultural life world is preexisting and is given to us by ancestors and fellow men along with a "ready-made standardized scheme ... as an unquestioned and unquestionable guide in all situations which normally occur within the social world" (Schütz 1970, p. 81). These schemes are passed down to us to supplement our own treasure trove of experience. Moreover, they serve as recipes for how to handle situations, carry out actions, and interpret the socially constituted life world. This handed down sociocultural knowledge is commonly given and accepted not only by the individual being, but by the fellow men as well (ibid., pp. 190, 82). Schütz continues:

> The world of my daily life is by no means my private world but is from the outset an intersubjective one, shared with my fellow men, experienced and interpreted by others: in brief, it is a world common to us all. (ibid., p. 163)

Hence, social interaction and the we-relationship are presupposed and a—if not the most—natural part of our common sense and life world. We adopt this so-called "natural attitude" (ibid., p. 183) towards the specific cultural and historical realities of sociality. This includes our being in this world, the corporeal existence of fellow men, as well as "their conscious life, the possibility of intercommunication" (ibid., p. 164), and interactional relationships. The natural attitude is not only directed towards fellow human beings, but also towards everything that is part of our life world and can be experienced by man. Our natural attitude is even extended to animals as well as inanimate objects (ibid., p. 169). This so-called "universal projection" (Luckmann 1983) is the reason for the human tendency to anthropomorphize robots, if they resemble us. Resemblance is achieved by the display of social cues and body languages. To observe someone conducting himself and interpreting the behavior is the first source of understanding and as Alfred Schütz notes, "[t]o a certain extent, it is sufficient for many practical purposes that I understand ... behavior." The body serves as a "field of expression" (Schütz 1991, p. 153)[4] for expressive movements and expressive actions. To make it clear: Expressive actions are actions that are carried out to "project content into the outer world and announce oneself to someone else (ibid., p. 162). An expressive action always has an underlying motive configured by consciousness. Expressive movements, on the other hand, are mien, gestures, and, in general, social cues that can be perceived and interpreted. Emotions or a specific experience are automatically assigned to the expressive movements I perceive.

However, what one person interprets into the expression movements or observed external events, what they signify for one person does not necessarily have the same meaning for the other. Only the observer experiences them as a signifier of the other's experiences and it remains doubtful whether they were "intentionally a sign or an expressive movement without communication intent" (ibid., p. 164). From a mere observation—"without communicational intent" it is, generally, not clear whether someone is conducting himself or acting. Conclusions are drawn from the way someone conducts himself with respect to his actions. The difference in those terms lies in the thinking process preceding the "observed, external event".

Schütz uses an analogy of wood cutting to explain the role of the observer and how we understand and bestow meaning onto the observed. He differentiates between three levels of understanding and what wood cutting can mean for the observer. The first two levels make statements about the "facticity of the situation" and the bodily movements sui generis without taking into consideration the

---

4   Quotes from Schütz (ibid.) are translated by the author.

"lived experiences" (Schütz 1970, p. 172) and cognitive processes underlying human actions.

1) In the first instance, we only consider the event of wood cutting. Having a look at the execution of cutting, we find that cutting may be done by a machine or a man. This is due to the fact that the procedure is identified as wood cutting by connecting it with the overall context of our experience and knowledge.

2) The second level of understanding concerns the bodily movements and the "observer's own perceptual experience" (ibid.). Every change and movement of the observed body is taken as indicator of its vitality and consciousness. Nothing else is assumed about the action, the underlying motives, and thinking processes.

Consequently, the understanding of an expressive movement depends on a spatio-temporal vicinity and the ability to compare the observed with something the observer is familiar with. However, human visions of the expressive movements observed are always connected with the automatically executed cognitive processes of perception. For the role of the observer, this means that the observation can never be reduced to the use of our visual capabilities, but we automatically perceive the observed as something; in the first case, as wood cutting and in the second case, as wood cutting by someone/something.

3) It is essential to note that the described observations of wood cutting are distant and detached observations. Still, in a third step, the observer takes the bodily movements as indicators of the body's vitality and consciousness, but also as indications of a specific conduct or an action. Schütz notes that the observer does not only focus on the indications as such, but also on the "what for." This means that the observer asks himself what the indications indicate with respect to the supposed experience of the observed. As Schütz puts it:

> I may, as I watch him, take my own perception of his body as signs of his conscious experiences. In so doing, I will take his movements, words, and so forth, into account as evidence. I will direct my attention to the subjective rather than the objective meaning-contexts of the indications I perceive. As a direct observer I can thus in one glance take in both the outward manifestations—or "products"- and the process in which are constituted the conscious experience lying behind them. This is possible because the lived experiences of the other are occurring simultaneously with my own objective interpretations of his words and gestures (ibid., p. 196).

We tend to look behind the surface behavior and try to make sense of it by trying to understand the motives[5] for the behavior and actions, which eventually results in a "genuine understanding of the other." Understanding the "outward manifestations"

---

5   Schütz (2011, p. 121) distinguishes two kinds of motives: *in-order-to* and *because-motives*. He defines them as follows: "Whereas the in-order-to relevance motivationally emanates from the already established paramount project, the because relevance deals with the motivation for the establishment of the paramount project itself."

and the "conscious experience lying behind them" is rooted in familiar structures and patterns that have been perceived previously, experienced, and typed into categories which serve as frame of reference or—in other words—act as tools to "survive in the social life world".

Understanding is not only characterized by the generalizing typification mentioned above, but also by a biographically determined situation. Everything we observe, we interpret by integrating it into the realm of our own experiences. This is because only our ego has a privileged access to ourselves and accesses ourselves through reflexive self-interpretation. We are able to look back on a past that shapes future actions: We gather a "stock of knowledge at hand" which forms "a scheme of interpretation of our past and present experience and also determines our anticipations of things to come" (Schütz 1970, p. 74). Thus, the stock of knowledge constitutes and contextualizes every understanding.

Transferring the explanations of observations above to an interaction situation, we can conclude that while we engage in an interaction, we simultaneously observe, understand, and interpret a situation. Everything we perceive of the other, e.g. social cues, is put into the context of our overall experience to anticipate the expectations of the other. Furthermore, interactions in a sense consist of either meeting or disappointing expectations. The anticipations and also the motifs we impute to the interaction partner influence how we will behave towards the other. Humans acquire their stock of knowledge and learn these "embodied practices" (Gallagher 2013, p. 557) early on and the latter serve as a mechanism to facilitate life and interactions as such.

A robot that partakes in our daily lives and is to function as a companion on level terms does not only need to be able to display believable social cues, but must be capable of recognizing them and reacting appropriately. However, this is not only restricted to the aspects of communication (including emotions), but also includes the array of situational, personal, cultural, historical contexts as well. If these contexts are arranged in a scheme of typification and included into the robot software, which would require a huge set of data, the robot might indeed be able to recognize surface behavior and use the human bodily movement as well as situational typifications as indicators, since, as Schütz outlines:

> [t]he sum-total of these various typifications constitutes a frame of reference in terms of which not only the sociocultural, but also the physical world has to be interpreted, a frame of reference that, in spite of its inconsistencies and its inherent opaqueness, is nonetheless sufficiently integrated and transparent to be used for solving most of the practical problems at hand. (Schütz 1970, p. 119)

The robot could assume the function of a detached observer, who translates and equilibrates the observed behavior with input behavioral patterns. If the robot is constructed for a specific purpose or role, the practical problems at hand can be defined as a situation in which the robot is supposed to display a certain behavior. What needs to be done would be a compendium of social situations a companion robot could be confronted with. These social situations need to be classified as typifications of appropriate social behavior, e.g. a robot that sees a crying human could react by hugging him/her ("comforting"). However, human behavior and needs are always more complicated. If we take crying as an example: We cry out of pain, sadness or joy. If we miscalculated the falling direction of a cut tree and got stuck under it, we do not need a hug. Although a hug because of sadness or even joyful exuberance might be appropriate, it may be inappropriate depending on the roles, e.g. student-teacher. In accordance with Breazeal (2002, p. 9), emulating human social interaction requires a robot to be able to identify who the person is (identification), what the person is doing (recognition), and how the person is doing it (emotive expression). While these demands might be technically feasible, Breazeal (ibid.) also claims that a robot equipped with these abilities can use such information to treat the person as an individual, to understand the person's surface behavior. For social interactions, however, the question why a person is doing something is essential and whether we can refer to him or at least understand his motives. We take it for granted that other fellow men can refer to us and that our perspectives are interchangeable. This interchangeability makes a true, reciprocal social interaction possible. The social life world and the recognition of the fellow men are based on a we-relationship between spatio-temporally coexisting subjects that experience each other. A true we-relationship is not only grounded on the acquisition of observed characteristics and the knowledge that the other is turned towards me, but rather on how he is turned towards me.

In a more distant social interaction social typifications and recipes might be applied to enable a regulated social interaction. But in a close social interaction, I do not only perceive my fellow men as another person, but as an individual to whom I refer in a second-person perspective. In the end, social interaction is what bestows meaning on life. That does not necessarily mean that a human cannot ascribe meaning to an interaction, or casually speaking, engage in a meaningful interaction. A human experiences his actions in this world as meaningful as such. For example, working with a robot can be meaningful in the way that a human perceives the work as important and part of this work might be that a robot helps to perform the work-related tasks. We can interact with robots in a work environment or even react towards emotions they display. Dautenhahn (1998) states in the context of socially intelligent agents (SIAs): "The process of building SIAs is influenced by what the human as the designer considers "social" and conversely, agent tools that are behaving socially can influence human conception of sociality" (ibid. p. 573). The question remains, if we want to head in this direction for better or worse. Because in the end, one important part of our life, world is, as the German phenomenologist Schütz states, that the interaction partner and I share the same experience. The experiences we make in our daily lives are very unique and shape who we are. It seems that a robot can have a certain function or a specific role, but human beings take over very different roles in their lives. And those roles are only a small part of who we are. In a true and meaningful we-relation these roles are recognized, but they do not define us as a human being.

Conclusion

This chapter shows that understanding always needs higher cognitive capacities in order to draw conclusions from the observed behavior of another being. A social interaction consists of more than a mutual observation. The interaction partners have to be aware of the other and his (biographically determined) situation and they have to have the ability to react to someone. A robot might have the ability to respond to me and recognize behavioral patterns, but a true social interaction will not be possible, unless I can see myself in the other and I know the other can take my perspective and will be able to understand me. Understanding always means to be able to identify the motives underlying an action and to transfer them into my own motives to achieve something together.

References

Breazeal, C. 2002. *Designing Sociable Robots*, edited by Ronald C. Arkin, Intelligent Robotics and Autonomous Agents Series. Cambridge, MA: A Bradford book.

Breazeal, C. 2004. "Social Interactions in HRI: The Robot View." *IEEE Transactions on Systems, Man, and Cybernetics Part C Applications and Reviews* 34(2): 181-6. doi: http://dx.doi.org/10.1109/Tsmcc.2004.826268.

Chin, K-Y., C-H. Wu, and Z-W. Hong. 2011. "A Humanoid Robot as a Teaching Assistant for Primary Education." 2011 Fifth International Conference on Genetic and Evolutionary Computing (ICGEC). 21-4. doi: http://dx.doi.org/10.1109/ICGEC.2011.13.

Dautenhahn, K. 1995. "Getting to Know Each Other: Artificial Social Intelligence for Autonomous Robots." *Robotics and Autonomous Systems* 16(2-4): 333-56. doi: http://dx.doi.org/10.1016/0921-8890(95)00054-2.

Dautenhahn, K. 1998. "The Art of Designing Socially Intelligent Agents: Science, Fiction, and the Human in the Loop." *Applied Artificial Intelligence* 12(7-8): 573-617. doi: http://dx.doi.org/10.1080/088395198117550.

Dautenhahn, K. 2007. "Socially Intelligent Robots: Dimensions of Human–Robot Interaction." *Philosophical Transactions of the Royal Society of London. Series B, Biological Sciences* 362(1480): 679-704. doi: http://dx.doi.org/10.1098/rstb.2006.2004.

Dourish, P. 2001. *Where the Action Is: The Foundation of Embodied Interaction*. Cambridge, MA: MIT Press.

Dreyfus, H.L. 1972. *What Computers Can't Do: A Critique of Artificial Reason*. New York: Harper & Row.

Faber, F., M. Bennewitz, C. Eppner, A. Gorog, C. Gonsior, D. Joho, M. Schreiber, and S. Behnke. 2009. "The Humanoid Museum Tour Guide Robotinho." The 18th IEEE International Symposium on Robot and Human Interactive

Communication, 2009. RO-MAN 2009. 891-6. doi: http://dx.doi.org/10.1109/ROMAN.2009.5326326.

Fong, T., I. Nourbakhsh, and K. Dautenhahn. 2003. "A Survey of Socially Interactive Robots." *Robotics and Autonomous Systems* 42(3-4): 143-66. doi: http://dx.doi.org/10.1016/S0921-8890(02)00372-X.

Gallagher, S. 2001. "The Practice of Mind. Theory, Simulation or Primary Interaction." *Journal of Consciousness* 8(5-7): 83-108. doi: http://dx.doi.org/10.1197/jamia.M1511.Database.

Gallagher, S. 2013. "You and I, Robot." *AI & Society* 28(4): 455-60. doi: http://dx.doi.org/10.1007/s00146-012-0420-4.

Giuliani, M., R.P.A. Petrick, M.E. Foster, A. Gaschler, A. Isard, M. Pateraki, and M. Sigalas. 2013. "Comparing Task-Based and Socially Intelligent Behaviour in a Robot Bartender." Proceedings of the 15th ACM on International conference on multimodal interaction - ICMI '13. 263-70. doi: http://dx.doi.org/10.1145/2522848.2522869.

Goodrich, M.A. and A.C. Schulz. 2007. "Human–Robot Interaction: A Survey." *Foundations and Trends in Human–Computer Interaction* 1(3): 203-75. doi: http://dx.doi.org/10.1561/1100000005.

Grathoff, R. 1989. *Milieu und Lebenswelt. Einführung in die Phänomenologische Soziologie und die sozialphänomenologische Forschung.* Frankfurt am Main: Suhrkamp.

Iacono, I., H. Lehmann, P. Marti, B. Robins, and K. Dautenhahn. 2011. "Robots as Social Mediators for Children with Autism - a Preliminary Analysis Comparing Two Different Robotic Platforms." 2011 IEEE International Conference on Development and Learning (ICDL). 1-6. doi: http://dx.doi.org/10.1109/DEVLRN.2011.6037322.

Ikemoto, S., T. Minato, and H. Ishiguro. 2008. "Analysis of Physical Human–Robot Interaction for Motor Learning with Physical Help." *2008 8th IEEE-RAS International Conference on Humanoid Robots (Humanoids 2008)*: 67-72. doi: http://dx.doi.org/10.1109/ICHR.2008.4755933.

Janich, P. 2012. "Between Innovative Forms of Technology and Human Autonomy: Possibilities and Limitations of the Technical Substitution of Human Work." In *Robo- and Informationethics: Some Fundamentals*, edited by Michael Decker and Mathias Gutmann, 211-30. Münster: LIT Verlag.

Kanda, T., R. Sato, N. Saiwaki, and H. Ishiguro. 2007. "A Two-Month Field Trial in an Elementary School for Long-Term Human Robot Interaction." *IEEE Transactions on Robotics* 23(5): 962-71. doi: http://dx.doi.org/10.1109/TRO.2007.904904.

Kosuge, K. and Y. Hirata. 2004. "Human–Robot-Interaction." Proceedings of the 2004 IEEE International Conference on Robotics and Biomimetics, August 22-26 2004, Shenyang, China. 8-11. doi: http://dx.doi.org/10.1109/ROBIO.2004.1521743.

Luckmann, T. 1983. "On Boundaries of the Social World." In *Life-World and Social Realites*, 40-67. Portsmouth: Heinemann.

Nieuwenhuisen, M. and S. Behnke. 2013. "Human-Like Interaction Skills for the Mobile Communication Robot Robotinho." *International Journal of Social Robotics. Special Issue on Emotional Expression and Its Applications* 5(4): 549-61. doi: http://dx.doi.org/10.1007/s12369-013-0206-y.

Scholtz, J. 2003. "Theory and Evaluation of Human Robot Interactions." System Sciences, 2003. Proceedings of the 36th Annual Hawaii International Conference on on System Sciences (HICSS'03). 10 pp. doi: http://dx.doi.org/10.1109/HICSS.2003.1174284.

Schütz, A. 1970. *On Phenomenology and Social Relations. Selected Writings*, edited by Helmut R. Wagner. Chicago, IL: The University of Chicago Press.

Schütz, A. 1991. *Der sinnhafte Aufbau der sozialen Welt. Eine Einleitung in die verstehende Soziologie.* 5 ed. Frankfurt am Main: Suhrkamp.

Schütz, A. 2011. "Reflections on the Problem of Relevance." In *Collected Papers V. Phenomenology and the Social Sciences*, edited by Lester Embree. Dodrecht: Springer Netherlands. doi: http://dx.doi.org/10.1007/978-94-007-1515-8.

Searle, J.R. 1980. "Minds, Brains, and Programs." *The Behavioral and Brain Sciences* 3(3): 417-57. doi: http://dx.doi.org/10.1017/S0140525X00005756.

Sekmen, A. and P. Challa. 2013. "Assessment of Adaptive Human–Robot Interactions." *Knowledge-Based Systems* 42: 49-59. doi: http://dx.doi.org/10.1016/j.knosys.2013.01.003.

Staudigl, M. 2014. "Reflections on the Relationship of "Social Phenomenology" and Hermeneutics in Alfred Schutz. An Introduction." In *Schutzian Phenomenology and Hermeneutic Traditions*, edited by Michael Staudigl and George Berguno, 1-6. Dodrecht: Springer Netherlands. doi: http://dx.doi.org/10.1007/978-94-007-6034-9_1.

Wang, H. and K. Kosuge. 2012. "Understanding and Reproducing Waltz Dancers' Body Dynamics in Physical Human–Robot Interaction." Robotics and Automation (ICRA), 2012 IEEE International Conference on, 14-18 May 2012. 3134-40. doi: http://dx.doi.org/10.1109/ICRA.2012.6224862.

Weber, J. 2014. "Opacity Versus Transparency. Modelling Human–Robot Interaction in Personal Service Robotics." *Science, Technology & Innovation* 10: 187-99.

Zhao, S.Y. 2006. "Humanoid Social Robots as a Medium of Communication." *New Media & Society* 8(3): 401-19. doi: http://dx.doi.org/10.1177/1461444806061951.

Chapter 2

Making Sense of Empathy with Sociable Robots: A New Look at the "Imaginative Perception of Emotion"

Josh Redstone[1]

In the field of social robotics, empathy is somewhat of a "hot topic" for engineers and empirical researchers alike. In this chapter, I examine a philosophical contribution toward making sense of empathy with robots, namely: Misselhorn's (2009) proposal that empathy with robots occurs through the interplay between perception and imagination ("imaginative perception"). I argue that although Misselhorn's explanatory framework captures something true about why people feel empathy for robots, such emotional responses are better conceived of as analogous to perceptual illusions rather than as cases of imaginative perception. I subsequently modify Misselhorn's framework to accommodate this, and then explore whether this modified framework can help account for another emotional response toward robots, the uncanny valley phenomenon. Here, I draw from some examples of empirical research on the uncanny valley. I show that people can (mis)perceive that robots possess attributes like animacy, emotion, and mentation. If one also believes that robots lack these things, then the inconsistency between that belief and the aforementioned misperceptions can make a robot seem eerie. Finally, I conclude that examples of people feeling empathy for robots are not genuine cases of empathy.

1   I wish to thank my co-supervisor David Matheson, along with my colleagues Benjamin James, Sara Grainger, and the audience at the Empathy and Understanding session at the RoboPhilosophy 2014: Sociable Robots and the Future of Social Relations conference for their helpful comments. My gratitude also goes out to Heidi Maibom, who provided helpful feedback on this chapter, and who taught me about empathy. Gratitude also goes out to my co-supervisor Guy Lacroix, who directed me toward some helpful empirical literature on the uncanny valley. Special thanks go out to my undergraduate supervisor Wayne Borody, who first introduced me to Masahiro Mori's work on the uncanny valley, and who inspired me to combine two of my passions—robots and philosophy—into an ongoing research project. Finally, thank you Marco Nørskov, for all of your hard work on this volume.

Introduction

Socialrobotics pioneer Cynthia Breazeal writes that a sociable robot is "socially intelligent in a humanlike way, and interacting with it is like interacting with another person." In other words, the essential idea behind the sociable robot is that it can communicate with, interact with, and understand people in the same "social terms" as people communicate with, interact with, and understand one another. Moreover, a sociable robot must also understand itself on these social terms. Such terms, according to Breazeal, include relating to the robot, or empathizing with it. "At the pinnacle of development," continues Breazeal, "[sociable robots] could befriend us, as we could them" (2002, p. 1).

In this chapter, I shall focus on one of these social terms, specifically empathy. Considering the human tendency to anthropomorphize, together with the humanlike social behaviors and emotive facial expressions that sociable robots are endowed with, it is understandable that sociable robots can facilitate comfortable humanrobot interaction and encourage those who interact with these robots to develop emotional bonds with them (cf. Breazeal 2002, 2003a, b). Those of us who have interacted with sociable robots have firsthand experience of just how contagious the emotive expressions of such artifacts can be. As it happens, empirical researchers have now begun to explore the behavioral basis of empathy with robots, and some of this research suggests that empathic responses are sometimes shown toward robots even if they are not altogether very humanlike. For example, one study found that people who observed the mistreatment of a robotic toy dinosaur called Pleo showed stronger physiological arousal than the participants who observed Pleo being treated well. These experimental participants also reported experiencing negative feelings, and felt "empathic concern" for the robot (Rosenthal-von der Pütten et al. 2012). In a similar vein, researchers have begun to explore the neurological basis for empathy with robots using fMRI imaging (e.g. Rosenthal-von der Pütten et al. 2013).

Cases where people show empathic responses toward robots are certainly compelling. Indeed, it would be rather strange if sociable robots did not elicit empathic responses from people who interact with them, for that is precisely the sort of response that engineers like Breazeal aim for their robots to elicit. However, considering that sociable robots are not capable of experiencing emotions (at least, not yet), an interesting conceptual problem emerges here: does it make sense to conceive of people's emotional responses toward sociable robots in terms of empathy? In other words, when people feel empathy for robots, does this *really* count as empathy? The first to notice this problem (so far as I am aware) is philosopher Catrin Misselhorn (2009), who not only draws attention to this problem, but also outlines a conceptual framework that explains how empathy with sociable robots and other humanlike artifacts is possible. She proposes that empathy with robots occurs as the result of the interplay between perception and imagination, which she calls "the imaginative perception of emotion." In other words, she thinks that people feel empathy for robots because they "imaginatively perceive" them to feel, and to express, emotions when in fact robots really cannot do either of these things. One of the aims of this chapter is to consider whether or not this imaginative perception framework can help to make sense of empathy with sociable robots, and if not, to modify it so that it can. [2]

To begin, I set the context for this discussion by making the aforementioned conceptual problem a little more concrete, using an illustrative example from science fiction. I then introduce and explain another noteworthy emotional response toward robots for which Misselhorn attempts to account with her imaginative perception framework, namely, the uncanny valley phenomenon. Then, I examine Misselhorn's conceptual framework in detail. I attempt to show that while her account captures something correct about why people report feeling empathy for robots, it might be conceptually clearer to characterize such phenomena as analogous to perceptual illusions, rather than as cases of imaginative perception. Since Misselhorn's other aim is to explain the uncanny valley phenomenon, I shall also explore whether the modified version of her framework that I defend here can do the same. While undertaking this analysis I shall consider not just emotional or affective varieties of empathy, but cognitive empathy or "theory of mind" as well. Finally, I shall answer the question of whether it makes sense to conceive of emotional responses toward sociable robots in terms of empathy. But let us begin by exploring the conceptual issue concerning empathy with sociable robots in greater detail.

From Empathy to Eeriness

The film *Robot & Frank* (2012) tells the story of Frank, a retiree in the near future who lives alone and shows signs of dementia. Concerned for his father's wellbeing, Frank's son, Hunter, gives him a domestic robot that provides companionship, cognitive stimulation, and encourages physical activities such as walking and gardening. Frank initially regards the robot as somewhat of a nuisance rather than as a companion. Nevertheless, it soon occurs to Frank that he can use his robot to restart his former career as a cat burglar. After the duo pulls a series of heists and attracts the attention of the local police, Frank is faced with a difficult decision. He must decide whether or not to delete his robot's memory in order to destroy the evidence of his crimes and to protect himself and his family from the consequences of his behavior. At this point Frank has begun to treat the robot more like a person—even a friend—than like a machine, and he becomes increasingly hesitant to delete its memory. By the time the police come knocking at his door to apprehend him, Frank has become so distressed by the decision he is faced with that the robot is forced to remind him that it is "not a person, just an advanced

---

2 I considered this question, albeit in somewhat less depth, in Redstone (2014). In what follows I shall attempt to further hone some of these ideas, and to discuss some of my previous work on the uncanny valley phenomenon.

simulation," and that by having Frank wipe its memory it will still have fulfilled its ultimate purpose, which is to help Frank.

I think that this example serves to make the conceptual problem identified by Misselhorn (2009) a bit more concrete. That is, if the humanlike characteristics of Frank's robot are simply an "advanced simulation"—in other words, if Frank's robot is not a real person capable of having real experiences, but merely a simulated person—then does it make sense to say that Frank's emotional experience for the robot, who he feels as though he is killing by deleting its memory, truly counts as a case of sympathy, compassion, or some other empathic response? The same question can be asked about the experiences of the experimental participants in the previously mentioned empirical studies (Rosenthal-von der Pütten et al. 2012, Rosenthal-von der Pütten et al. 2013). Surely these people felt some sort of sympathic response toward the mistreatment of Pleo. But if Pleo is not capable of having emotional experiences, or of feeling pain, discomfort, etc., do these peoples' experiences toward Pleo truly count as cases of sympathy? And, do other ostensibly empathic responses toward humanlike robots really count as cases of empathy?

As I mentioned in the introduction, Misselhorn (2009) goes further than simply identifying this as an interesting philosophical problem. She also provides a framework that, in her words, "[explains] how empathy with inanimate objects is possible" (ibid., p. 356).[3] Specifically, she points out that people do not literally perceive emotional expressions when they interact with artifacts like sociable robots because such artifacts do not feel, and therefore cannot literally express, emotions.[4] Misselhorn therefore posits that the imagination is somehow involved when people feel empathy for robots, a phenomenon that she calls the "imaginative perception of emotion." I shall explain how Misselhorn thinks that the imaginative perception of emotion works below. First, however, there is another important aspect of Misselhorn's conceptual framework which requires some attention. That is, understanding empathy with inanimate objects is, for Misselhorn, a step toward explaining another noteworthy emotional response toward robots: the socalled "uncanny valley" phenomenon. Since explaining this phenomenon is one of Misselhorn's principle aims, let us briefly discuss the uncanny valley in order to place her framework for empathy with inanimate objects into context.

The uncanny valley phenomenon was described by roboticist Masahiro Mori (2012). He observes that humanlike artifacts, such as toy robots and prosthetic limbs, can elicit a sense of familiarity, or affinity,[5] so long as their appearance and

---

3 Since sociable robots can move about, make facial expressions, etc., they are certainly not one's typical inanimate objects. I return to this observation toward the end of this chapter.

4 For similar discussion, see Dennett (1981).

5 The Japanese word that Mori (2012) uses, *shinwakan*, has been variously translated by various translators. It sometimes appears as "familiarity," and sometimes as "affinity." Bartneck et al. (2009) suggest that a more appropriate rendering is "likeability," since

movement is only somewhat humanlike. However, if such an artifact were to look and move a lot like the human equivalent—for example, an artificial hand covered with silicon skin that feels cold to the touch—then it can stop eliciting a sense of affinity and instead elicit a sense of eeriness. The aforementioned artificial hand, Mori observes, would become uncanny. Extrapolating from these observations, he hypothesizes that "[s]ince the negative effects of movement are apparent even with a prosthetic hand, to build a whole robot would magnify the creepiness" (ibid., p. 100).

Mori's (2012) uncanny valley hypothesis, then, is as follows: a humanlike artifact such as a robot whose physical appearance and movement is more robotic, i.e. mechanical, than humanlike can elicit a sense of affinity or "likeability." Furthermore, the more humanlike the robot's appearance and movement are, the stronger one's sense of affinity toward it will be. However, Mori posits that a very humanlike robot—such as an android with silicon skin, hair, facial expressions, etc.—can fail to elicit a positive emotional response and instead elicit the a sense of eeriness. When represented graphically, with humanlikeness on the X axis and affinity or likeability on the Y axis, this shift from affinity to a sense of eeriness appears as a U-shaped curve—thus, the uncanny valley phenomenon.

Note that, as historian Minsoo Kang (2011) points out, at the time Mori made the observations which would lead him to formulate his uncanny valley hypothesis, "no robot existed whose resemblance to a human being was so perfect that one could verify whether it did pull itself out of the uncanny valley" (ibid., p. 47). Indeed, Mori's (2012) hypothesis had to wait nearly 40 years before it finally caught the attention of empirical researchers who possessed the means with which to test it, although fortunately such empirical work is now well underway, as I discuss later on.

One additional thing to note here is that Misselhorn conceives of the range of emotional responses shown toward humanlike artifacts not in terms of affinity or likeability and eeriness, but in terms of *empathy* and eeriness. That is, what Mori (ibid.) refers to using the Japanese word *shinwakan*, Misselhorn (2009) refers to as empathy. The sense of eeriness one experiences in the uncanny valley phenomenon, she contends, results from the failure of empathy for the artifact, which in turn is a result of the failure of imaginative perception. Obviously, before Misselhorn can formulate an explanation for the uncanny valley phenomenon along the lines described above, she requires a way of conceptualizing empathy for artifacts which, she writes, do not "really have emotions nor … literally show emotional expressions" (ibid., p. 352). This is the impetus behind her imaginative perception framework. Now, before moving on to discuss imaginative perception itself, let us ground this discussion further by outlining a few examples of the kinds of emotional and cognitive phenomena that are called empathy.

---

familiarity can change over time and with repeated exposure to a stimulus. Here, I must defer to their knowledge of the Japanese language.

On Empathy

For many people, "empathy"—as the common expression goes—refers to putting oneself in another's shoes (cf. Misselhorn 2009, Maibom 2007, Darwall 1998, Goldman 1989). Of course, as philosopher Heidi Maibom (2012) points out, the terminology surrounding empathy is not always very straightforward. Sometimes altogether different terms are used to refer to the very same, or similar, emotional resonance phenomena. For example, "sympathy" is sometimes referred to as "empathic concern" (e.g. Batson 2011). As we have seen, Rosenthal-von der Pütten et al. (2012) and Rosenthal-von der Pütten et al. (2013) use "empathic concern" where I use "sympathy." But without getting ahead of ourselves, an important thing to note is that it is common to distinguish between affective varieties of empathy and cognitive empathy. In what follows I shall outline a few of these varieties, drawing primarily from Maibom's (2012, 2014) excellent discussion of empathy.

Let us begin with the affective varieties of empathy. According to Maibom (ibid.), one experiences affective empathy when one has an emotional experience that is consonant with, and similarly valenced to, the emotional experience of another. Moreover, one feels the emotion not for oneself, but for the other person in his or her situation. For example, let us say that a colleague of mine has just been accepted into graduate school at a prestigious university, and is very happy about the prospect of attending such an institution. If upon hearing this news I feel happy for my colleague, then I have empathized with her. Note, however, that it is only if I feel the emotion *for her*—that is, I am happy for her and not simply happy—that my feeling happy counts as empathy.

Sympathy is an emotional response which, according to Maibom (ibid.), is similar to empathy, but with an important difference. The difference is that while empathy is an emotional experience that is consonant with another's, sympathy is an emotional reaction toward another's welfare. So, when someone feels sympathy, one's emotional experience might be similarly valenced, although not necessarily identical, to the emotional experience of the other. Like empathy, however, sympathy is felt for the other person, and not for oneself. For example, imagine that my aforementioned colleague were rejected from rather than accepted into graduate school. She might feel sad about her situation. Then again, she might not feel particularly sad at all, having decided that graduate school might not be for her after all. However she feels, if I were to feel bad for her—knowing how bright she is and how much she might have enjoyed graduate school, say—and, if I were to also feel concerned about her in her situation, then I will have sympathized with her.

Two other emotional resonance phenomena that are related to empathy and sympathy are emotional contagion and personal distress. According to Maibom (ibid.), emotional contagion refers to when one "catches" the emotions of another. Think here of the last time you felt the excitement of the crowd at a rock concert or a football match. Or, think of those times when you automatically return a smile to someone who smiles at you. These are examples of emotional contagion. Moreover, unlike empathy and sympathy, when one experiences emotional contagion, the emotion is felt for oneself and not for the other. Personal distress, on the other hand, refers to the negative emotional experience felt upon encountering another in distress. Similar to emotional contagion, during personal distress the emotion is felt for oneself rather than for another. Finally, there is cognitive empathy, which is often called "theory of mind." Theory of mind concerns understanding the thoughts, beliefs, desires, etc., of other people. Since at this point in this chapter I am primarily interested in emotional responses toward sociable robots, I will confine my discussion to affective varieties of empathy for the time being, and return to the topic of cognitive empathy later on.

On the Imaginative Perception of Emotion

I can now explain exactly how Misselhorn (2009) answers the question: how is empathy with inanimate objects possible? Misselhorn takes one of Maibom's routes toward empathy as a starting point. According to Maibom, "S empathizes with O's experience of emotion E if S perceives O's T-ing [of E] and this perception causes S to feel E for O" (2007, p. 168).[6] But recall that Misselhorn is not committed to the view that robots really possess emotions. To accommodate this, she modifies Maibom's route to empathy, and obtains the following: "S empathizes with an inanimate object's imagined experience of emotion E if S imaginatively perceives the inanimate object's T-ing [of E] and this imaginative perception causes S to feel E for the inanimate object" (2009, pp. 351-2).

In what sense is imaginative perception imaginative? According to Misselhorn, the imagining that occurs during imaginative perception is not simply propositional. In other words, imaginative perception is not an active imagining of some particular thing, as is, for example, closing one's eyes and imagining a wingedhorse, or to use Misselhorn's example, imagining that a banana is a telephone. Instead, by "imaginative perception" Misselhorn seems to mean something that is much more automatic and unconscious than an active, intentional imagining. Specifically, she proposes that certain salient similarities between a perceived object (A), and another kind of object (B), causes concepts that are typically activated when perceiving a B to be activated when perceiving an A. She proposes further that the activation of these concepts can influence the "phenomenal feel" of the perceived object. However, on her view, this does not affect the content of the perception. In other words, one perceives an A, and one

---

6   Here, I am borrowing Maibom's notation, which Misselhorn (2009) also borrows: S refers to a subject; O refers to another person; E refers to an emotional experience; and finally, T-ing refers to the expression of an emotion. Note further that what I have outlined here is just one out of a number of routes to empathy discussed by Maibom. For further discussion, see Maibom (2007).

knows that one perceives an A, but owing to the similarity between A and B, the result of imaginative perception is that perceiving an A feels phenomenally similar to perceiving a B. Moreover, this feeling of similarity can become very strong if the relevant features of the perceived object are sufficiently numerous and salient, even though the perceiver knows she perceives one kind of thing (A) and not the other (B). In Misselhorn's words:

> The humanlike features M of an inanimate object trigger the concept of a human N, [and] for that reason, seeing the T-ing of an inanimate object feels (to some extent) like seeing a human T-ing, e.g., the humanlike features of a doll's face trigger the concept of a human face, [and] for that reason, seeing the smile of a doll feels like seeing the smile of a human being. Given that perceiving facial expressions can cause the same emotions in us … , we have by now arrived at the core of my explanation of how empathy with inanimate objects is possible. (2009, p. 354)[7]

Misselhorn maintains that the above process is a kind of imagining, specifying that the concept(s) N that are triggered by an object M are "entertained 'offline' … For this reason it is a kind of imagining, although not actively done" (ibid.). I am not entirely convinced that this process is really a kind of imagining, although I do think that there is something correct about Misselhorn's conceptual framework when it comes to explaining why people feel empathy for robots. Indeed, her suggestion that the humanlike features of a humanlike artifact can trigger humanassociated concepts and influence the "phenomenal feel" of perceiving, e.g., a sociable robot strikes me as quite plausible. I simply think that there might be a more terminologically precise way of describing this phenomenon besides imaginative perception. Specifically, it seems to me that what Misselhorn calls "imaginative perception" is better conceived of as analogous to a perceptual illusion—a misperception, if you like.

One reason why I am hesitant to characterize all of this as an imaginative process is that there is quite a bit of lexical flexibility surrounding words like "imaginative" or "imagination." For instance, sometimes we call things imaginary in the sense that they are pretend—such as Misselhorn's example of the bananaphone—and sometimes we call things imaginary which might be better described as misperceptions. Think here of a young child who becomes frightened after seeing what she thinks is a monster in her closet. Of course, there are no monsters in her closet; here we would say the monster is "just in her imagination." The child's parents might try to alleviate her fears by saying "It's just your imagination … see, there's no monster in here!" Rather, that which appears to the child in the darkness to be some sort of monstrous figure is revealed by the

---

7 Note, per my earlier discussion, that the phenomenon Misselhorn mentions here can be understood as emotional contagion. It is not clear to me that this would count as fullon empathy.

light of day to be a pile of clothing, boxes of toys, etc. Even though in everyday speech we might say that the monster existed only in the child's imagination, I think that it is more technically precise to characterize such an experience as a kind of misperception. Needless to say, I think that the same thing applies to the phenomena that Misselhorn would identify as cases of imaginative perception.

Here you are probably wondering: what difference does it make whether we treat the child's closetmonster as an imaginary monster or as a misperception of a monster? Even if one has no problem thinking of the above-described scenario in terms of an imagining, what reason is there to accept my alternative beyond the semantic ambiguity of words like "imagination"? One reason is that there are already imaginative routes to empathy that are well established in the empathy literature. These routes to empathy, unlike the imaginative perception of emotion, do involve an active imagining. Take for example another one of Maibom's (2007) routes to empathy, where "S empathizes with O's experience of emotion E in C, if S imagines being in C and imagining being in C causes S to feel E for O" (ibid., p 173).[8] This route is known as simulation. When one empathizes by simulating, one imaginatively projects oneself into another's situation, or imagines oneself to be the other, and then simulates (by imagining) what the other's emotional experience is like. The end result of a successful simulation, Maibom explains, is S's having an empathic experience for O in C. This is clearly quite different from the kind of imagining that Misselhorn (2009) has in mind. But since the above-described imaginative route to empathy is well established in the empathy literature—and since the precise meanings of words like "imaginative" or "imaginary" can be somewhat ambiguous—perhaps we ought to modify Misselhorn's framework accordingly.

I have already gestured toward one modification: instead of conceptualizing empathy for robots in terms of the imaginative perception of emotion, it can instead be thought of as analogous to perceptual illusions—or, to misperceptions. To illustrate the ways in which I think empathy for sociable robots and perceptual illusions are analogous, consider for example a visual illusion called the neon color spreading illusion. This illusion occurs when one observes a picture composed of mostly black lines on a white background. In a small part of the image—a circular portion in the center, say—the lines are colored rather than black. Upon observing a color spreading image, people report that the white background area over which the colored lines cross "glows" in a hue similar to that of the colored lines. Of course, upon close inspection, one can see that the background over which the colored lines cross is not colored at all. Yet, even when one is aware that this phenomenon is just an illusion—as with many other visual illusions, such as the Moon illusion or the Müller-Lyer illusion, or even in cases of auditory illusions

---

8   Here, C refers to some situation or circumstance.

such as the McGurk effect—one nonetheless continues to experience the color spreading effect when one looks at the image.[9]

The specific details concerning what mechanisms underlie the neon color spreading illusion are not well understood, and in any case they are beyond the scope of this chapter. These mechanisms are probably quite different than those that underlie our experiences of empathy for robots, but to complete this analogy, consider the following: when one experiences the color spreading illusion, one has the phenomenal experience of a colored background when, in reality, there is just a white background with colored lines crossing over it. Moreover, one still experiences the illusion even if one is aware that the background is not really colored. In other words, whatever perceptual mechanisms underlie this illusion, they are likely cognitively impenetrable. I propose that such perceptual illusions are analogous to cases where people feel empathy for sociable robots in a number of ways. Firstly, as Misselhorn (2009) and I would undoubtedly agree, one can experience feelings of empathy or sympathy for a robot even though the robot does not experience emotions itself. So, similar to how one has the phenomenal experience of the colored background in the color spreading illusion, one can feel empathy for a sociable robot owing to its emotive expressions and behaviors, even though the robot does not experience emotions. Secondly, even if one *knows* that the robot does not experience emotions, one might, for example, still cringe upon seeing the emotive expressions a robot makes when it is being mistreated. Thus, the perceptual processes underlying people's experiences of empathy for robots might be cognitively impenetrable as well. This is yet another reason why feeling empathy for robots is analogous to experiencing a perceptual illusion, in that even if one already knows, for example, that the background of a color spreading image is in fact not colored at all, but is simply a plain white background, one still experiences the illusion nonetheless.

I think that this perceptual illusion analogy is rather straightforward, and that it has the virtue of avoiding any potential confusion that might arise owing to the semantic ambiguity of words like "imaginary" and "imagination." It can also help to avoid any confusion that might arise owing to the established understanding of imaginative routes to empathy, such as empathy by simulation. Of course, I should emphasize that the conceptual framework I have presented above is more of an attempt at modifying Misselhorn's framework rather than an attempt at refuting it altogether. Again, I simply think that it serves as a candidate for more precisely characterizing what Misselhorn calls "imaginative perception" while still capturing how the perceptual processes she refers to using that term works. Now, recall that Misselhorn's aim, besides explaining how empathy for inanimate

---

9 Of course, a picture is worth a thousand words. I therefore recommend that the reader experience these illusions for him or herself. Michael Bach, a vision researcher at the University of Freiburg, keeps a website that serves as an excellent resource on the neon color spreading illusion, along with many other visual illusions: http://www.michaelbach. de/ot/colneon/index.html.

objects is possible, is to explain the uncanny valley phenomenon. I shall now consider whether the modifications I have made so far to Misselhorn's framework also prove helpful for achieving this, and whether there are any other ways that the framework can be modified or expanded in pursuit of this goal.

Empathy and the Uncanny Valley

Earlier I mentioned that Misselhorn's explanation of the uncanny valley phenomenon involves a failure of empathy—itself a failure of the imaginative perception of emotion—that leads to a sense of eeriness.[10] She describes this failure as follows:

> [B]ecause of the similarity of the features [of humanlike artifacts and human beings,] the [human] concept is triggered again and is repeatedly about to be elicited. This leads to a kind of very fast oscillation between four situations which resembles a gestalt switch:[11] the mere triggering of the concept, the reaching of the threshold of concept application, the failure of concept application resulting in a complete turning off of the concept, and the renewed triggering in keeping on to perceive the object. (2009, pp. 356-7)

One concern that might be raised here is this: why should a failure of empathy cause someone to experience a sense of eeriness in the presence of a sociable robot or some other humanlike artifact? Interestingly, there is some evidence that emotional responses toward humanlike artifacts can be mediated by the emotional displays of those artifacts, and in light of my earlier discussion of empathy and emotional contagion, this is not surprising. So admittedly, Misselhorn's suggestion that a failure of empathy might lie behind the uncanny valley phenomenon is not immediately problematic. For example, Tinwell et al. (2011) used computer-generated animations of human faces in order to see whether or not perceiving mismatched emotional expressions might have an effect on people's responses to those faces. The faces displayed each of the six emotional expressions common across all human cultures: fear, anger, disgust, surprise, sadness, and happiness, along with a neutral face as a control. One finding of this study was that the participants preferred real faces to computer-generated faces. But the most interesting finding, I think, is that the participants preferred the computer-generated faces even less when motion in the upper parts of the faces was limited. This also caused the participants some confusion when it came to determining what emotional expressions the limited motion faces were displaying. So, perhaps

---

10   Note that this section represents an attempt to further refine some of my previous work on the uncanny valley phenomenon (cf. Redstone 2013).

11   A gestalt switch occurs when a static image appears to change. The classic example is the Necker cube, a static 2D image whose orientation changes as an observer stares at it.

odd or mismatched emotional facial expressions might elicit a sense of eeriness, or at least they might cause those faces to be preferred less, i.e. become less likable than faces with emotive facial expressions that better mimic the human equivalent. If a robot were to possess the same sort of odd or mismatched facial expressions, then based on Tinwell et al.'s findings, we can easily imagine how empathy might be interfered with, which could make the robot seem very strange indeed.

Yet, as I discussed earlier, there are many different emotional and cognitive phenomena called "empathy." Moreover, sometimes different researchers refer to the same empathic phenomena using altogether different terms (cf. Maibom 2012, 2014). So ascertaining whether or not the failure of empathy accounts for the uncanny valley phenomenon might be made easier by clarifying what specific varieties of empathy are best to examine here. Recall that Misselhorn (2009) uses "empathy" in place of "affinity," "familiarity," and "likeability," the usual translations of *shinwakan*. Following my earlier overview of the varieties of empathy, however, it is clear that many of the examples of people feeling empathy for robots that Misselhorn discusses are cases of sympathy. Specifically, she mentions two empirical studies that used robots (Bartneck et al. 2005) and computer-generated animations of people (Slater et al. 2006) in Milgram-style obedience paradigms. In Bartneck's and his colleagues' (2005) study, for instance all 20 of the experimental participants administered what they were told was a lethal electric shock to a small toy robot, which responded to the simulated shocks as if it were suffering. Some of the participants reported that they felt sorry for the robot even though they knew it could not feel pain.[12]

These observations aside, I find it difficult to see exactly how a failure of sympathy or empathy would result in an uncanny sensation. But since my aim here is to be constructive rather than critical, I would like to suggest a way in which this can be addressed. Firstly I suggest that it is not a failure of empathy that causes one to experience a sense of eeriness when interacting with an uncanny humanlike artifact, but the *persistence* of empathy. Secondly, I think that this persistence can occur when it comes to affective varieties of empathy *and* cognitive empathy. To illustrate my case, consider Ernst Jentsch's (2008) observations of automata, the historical predecessors of modern robots. Jentsch's work is often mentioned in modern day examinations of the uncanny valley phenomenon, no doubt because his essay is arguably the first to explore the uncanny from a scientific perspective, and because of the uncanny similarity between his observations about automata and Mori's observations about humanlike robots. Specifically, Jentsch thinks that

---

12   As my colleague Benjamin James, along with some of the attendees of the RoboPhilosophy 2014 conference in Aarhus, Denmark have remarked to me, the idea of prompting people to "punish" other agents in these Milgram-style paradigms, be they humans or robots, is itself quite "creepy." James and I also wonder whether asking people to punish robots is ethical. For although robots do not feel pain, perhaps a robot's pain behavior could cause someone to experience personal distress, just as Frank experienced distress when he was forced to delete his robot's memory.

automata can seem quite uncanny when they "appear to be united with certain bodily or mental functions." He explains that:

> For many sensitive souls, [a lifesize wax or similar figure] also has the ability to retain its unpleasantness after the individual has taken a decision as to whether it is animate or not. Here it is probably a matter of semiconscious secondary doubts which are repeatedly and automatically aroused anew when one looks again and perceives the finer details; or perhaps it is also a mere matter of the lively recollection of the first awkward impression lingering in one's mind ...

Jentsch continues:

> This is where the impression easily produced by the automatic figures belongs that is so awkward for many people ... A doll which closes and opens its eyes by itself, or a small automatic toy, will cause no notable sensation of this kind, while on the other hand, for example, the lifesize machines that perform complicated tasks, blow trumpets, dance and so forth, very easily give one a feeling of unease. The finer the mechanism and the truer to nature the formal reproduction, the more strongly will this special effect make its appearance. (ibid., pp. 222-3)

Jentsch's observations suggest to me that it is not just affective empathy that lies behind the uncanny valley phenomenon. Instead, perhaps sociable robots elicit a sense of eeriness not only because they appear to feel emotions, but also because they appear as if they are alive, and that they can think. If this is correct, then I suggest that empathy—whether affective or cognitive—does not fail during the uncanny valley phenomenon. This proposal strikes me as compatible with a suggestion of Hiroshi Ishiguro's (2006, 2007), namely: that very humanlike robots might be recognized as different entities at different levels of cognitive processing. According to Ishiguro, it is possible that people recognize a robot as a robot on a conscious level while owing to the robot's humanlike appearance and behavior, on an unconscious level the robot is recognized as a human being. So, to restate some of Jentsch's (2008) observations in more modern terminology, owing to the automaticity and cognitive impenetrability of human perceptual systems, it is possible one *continues* rather than *fails* to perceive attributes like animacy, emotion, and mentation, even though one knows that robots are not really alive, do not have emotions, and do not have humanlike minds.[13]

I think that what I have sketched above is also compatible with the modified version of Misselhorn's (2009) framework for empathy with robots that I outlined

---

13 David Matheson has suggested to me that the knowledge that a humanlike artifact is not alive, together with the strong impression of animacy it elicits in an observer, might also give rise to a sense that someone is attempting to deceive the observer (perhaps even with a harmful intent). In turn, this could also elicit a sense of unease. I think that this suggestion is quite plausible, and that it is a potentially fruitful topic for further empirical study.

in the previous section. But admittedly, what I have presented thus far is a bit speculative, so I shall now outline some empirical findings that, I think, lend support to this modified framework. One such finding is that categorization difficulty seems to be involved in the uncanny valley phenomenon. In this case, categorization refers to assigning an entity membership in one category or another, e.g. 'animate' or 'inanimate,' 'human' or 'nothuman,' etc. In a paper by Burleigh, Schoenherr, and Lacroix (2013),[14] two experimental studies were conducted in order to investigate the uncanny valley phenomenon. The results of one of these studies suggest that categorization difficulty contributes toward a stimulus' ability to elicit a sense of eeriness. Burleigh and colleagues (ibid.) call this explanation of the uncanny valley phenomenon the "category conflict hypothesis." In this study, the participants' sense of eeriness was measured using 7-point Likert-scales. But in addition to investigating whether categorization difficulty might explain the uncanny valley phenomenon, Burleigh and colleagues also tested whether atypical features and levels of realism would cause the participants to rate the stimuli more harshly. Using computer-generated imagery, they created a between category continuum of images where a human face gradually morphed into a face with nonhuman animal features (such as a goatlike nose and ears), and a within category continuum of images where the level of realism of the faces was manipulated. In both of these experimental conditions, an atypical feature was also included in the series of images (such as one eye larger than the other). In the realism condition, it was found that the relationship between the participants' eeriness ratings and levels of realism of the images were linear, i.e. no uncanny valley was observed in the data. But interestingly, an uncanny valley effect was found when the participants rated the categorically ambiguous stimuli in the continuum of images that contained human and nonhuman animal features. Moreover, the participants rated images the most eerie when the category membership of images was the most ambiguous, at "approximately the midpoint of subjective human likeness" (2013, p. 770).

In a different series of studies conducted by Yamada, Kawabe, and Ihaya (2013), a technique called morphing was used to produce a series of images of one entity changing into another, such as a photograph of a human morphed with a cartoon image of a human and a photograph of a plush toy of a human. This series of morphed images was randomized and shown to the participants, who were asked to rate the likeability, rather than the eeriness, of the images using Likert-scales, and to assign them to one category or another (real human or cartoon human, etc.). Using the participants' reaction times as a measure of the categorical ambiguity of an image, it was found that it took the participants longer to categorize images that were categorically ambiguous, i.e. images that shared a lot of features from two

---

14   I would like to thank Guy Lacroix for bringing this excellent paper to my attention, and for his discussions with me about categorization difficulty and the uncanny valley phenomenon.

distinct categories. Interestingly, it was also found that the likeability ratings of the images were lowest when categorization latency was highest.

Arguably, both Burleigh et al.'s (2013) and Yamada et al.'s (2013) results suggest that negative evaluation occurs most strongly at what are, ostensibly, categorical boundaries. Burleigh et al. (2013) also suggest that the sense of eeriness elicited by the categorically ambiguous stimuli presented to their experimental participants can be understood in terms of an analogy with cognitive dissonance, i.e. the sense of psychological discomfort felt by someone who entertains two inconsistent beliefs simultaneously (cf. Festinger 1957). When it comes to uncanny robots, perhaps they elicit a sense of eeriness because one entertains a belief about robots (e.g., that they are inanimate) that is inconsistent with her perception of the robot (e.g., that it appears animate). In any case, I think that this analogy from Burleigh et al. (2013) dovetails nicely with Jentsch's (2008) remarks about uncanny automata and Ishiguro's (2006, 2007) suggestions concerning conscious and unconscious recognition of humanlike robots.

I should add that, while neither of these studies examined animacy, there are other empirical studies that have investigated when and for what reasons people make attributions of animacy to certain entities. The findings of one such study, conducted by Looser and Wheatley (2010), also complement what I have sketched in this section nicely. The authors used a series of morphed images of real human faces and humanlike mannequin faces in order to see whether they could determine the point on a continuum of humanlike appearance at which people begin to attribute animacy to a human face. They observed, across a number of different experimental conditions, that the participants attributed animacy to the stimuli at roughly the same level of humanlikeness at which they also made attributions of mind such as agency and experience (cf. Gray, Gray, and Wegner 2007). Looser and Wheatley (2010) also observed that the participants took the longest to categorize an image as either a real human or a mannequin at roughly the same point at which they began to make attributions of mind, and concluded that their participants made attributions of animacy and mind at a categorical boundary between the animate and the inanimate. It is also interesting to note that, even though this study did not investigate the uncanny valley phenomenon, Looser and Wheatley report that some of their experimental participants remarked that images near this categorical boundary seemed "creepy or unsettling" (ibid., p. 1860).

I think that another empirical study, which specifically investigated whether mind perception is involved in the uncanny valley phenomenon, lends plausibility to the framework I'm building here. Gray and Wegner (2012) ran two studies in which the levels of agency and experience of the stimuli presented to their experimental participants were manipulated. In one of these studies, they gave their participants one of two descriptions a computer: one group of participants was told the computer was simply a powerful supercomputer; the other group was told that the computer was so powerful that it could experience sensations like pain or thirst. In the other study, the participants were given a description of a man, accompanied by a picture. One group was told the man in the picture was "normal," another group that he lacked the capacity for experience, and the final group that he lacked agency. The different groups of participants were asked to rate the perceived "unnervingness" of this man and the supercomputer using Likert-scales. Gray and Wegner found that the participants rated the computer that was described to have experience as more unnerving than the computer that was simply described as a powerful supercomputer. Similarly, the participants who were told the man lacked the capacity for experience reported higher levels of unnervingness compared to those who were told he lacked agency or that he was otherwise normal. These results suggest that when the capacity for experience—one of the "attributes of mind" identified by Gray, Gray, and Wegner (2007)—seems present where people don't expect it (in the case of the supercomputer), or seems absent where people expect it to be present (in the case of the man), this can cause people to experience unease. Gray and Wegner conclude that because "higher cognition" has been regarded as a defining characteristic of the human being since antiquity, and that "because humans are fundamentally expected to have experience, a person perceived to lack experience ... should be seen as unnerving" (2012, p. 126).

Earlier I mentioned that we should specify other varieties of empathic experience besides sympathy as possible factors in the uncanny valley phenomenon. If the empirical evidence I have reviewed so far is any indication, then there is reason not to focus solely on affective empathy or sympathy, but to include cognitive empathy as well. However, the reader might raise a concern against what I have presented so far: perceiving attributes of mind might not count as fullon cognitive empathy. Perhaps this is the case, although I think that the idea that people perceive robots to exhibit intentional behavior—that is, to exhibit goaloriented behavior, as if they have minds and are somehow thinking about what they are doing—is a promising one. For example, Saygin et al. (2011) investigated the uncanny valley phenomenon used fMRI imaging to examine what happened when their experimental participants observed goaloriented robot behavior. Their experiment had three conditions, in which the participants observed a mechanical looking robot, an android, or a real human, performing gestures such as waving a hand as if to greet someone. Interestingly, it was found that the participants in the android condition had the highest levels of activity in the areas of the brain called the "actionperception system" (APS),[15] while the participants in the robot and human conditions experienced lower levels of neural activity in these areas. Saygin and colleagues concluded that the robot's and the human's appearance

---

15   The APS includes brain structures like the lateral temporal, anterior intraparietal, and inferior frontal/ventral premotor cortices, some of the same areas that also make up what some call the "mirror neuron system" (MNS). The MNS is thought to contain neurons which fire both when one performs an action and when one observes someone else performing the same action. According to Iacoboni (2009), mirror neurons might explain how people represent the contents of others' minds, thereby forming a neurological basis for empathy.

corresponded to what the APS "expected" the observed actions of the agent would be like. The reason the levels of APS activity were higher in the android condition, they suggest, owes to the fact that the android's biological appearance was inconsistent with its mechanical motions. This not only provides some insight into the neurological basis of the uncanny valley phenomenon, but also suggests that people understand robots as intentional agents rather than mindless automata. To borrow Dennett's (1989) terminology, it might be said that people adopt the Intentional Stance when it comes to understanding how robots behave.

Conclusion

My aim in this chapter has been to determine whether cases of people feeling empathy for robots count as genuine cases of empathy. My approach toward this question has been to examine Misselhorn's (2009) framework for explaining how empathy with robots is possible. I have attempted to show that while there is something correct about her account of empathy with robots—primarily, that the similarities between sociable robots and human beings can cause perceiving a robot to feel phenomenally similar to perceiving a human—it might be better to characterize this as analogous to a perceptual illusion rather than as imaginative perception. When it comes to Misselhorn's other aim, understanding the uncanny valley, I think that the misperceptioninspired modifications I have made to her framework leave it capable of integrating and explaining empirical research on this phenomenon as well.

Although I have aimed at being constructive rather than critical in this chapter, and although I think that Misselhorn's (2009) conceptual framework captures something correct about why people feel empathy for robots, there is one final issue with her framework that I must briefly discuss. That is, her account is directed toward explaining empathy with *inanimate objects*. But as I have shown, sociable robots are certainly not one's typical inanimate objects. Instead, sociable robots behave as if they have emotions, even though in actuality they do not. They also quite easily elicit perceptions—or misperceptions if you like—of animacy. Moreover, a robot's behaviors are often goaloriented, i.e. intentional, which can make them seem to an observer as though they have minds of their own. So, one very important reason why people feel empathy for these inanimate objects is that they often do not seem inanimate to an observer.

Finally, I must answer the question of whether or not cases of people feeling empathy for robots count as genuine cases of empathy. If one accepts Misselhorn's (2009) imaginative perception framework as it is, then it seems to me that feeling empathy for sociable robots does not count as genuine empathy. Of course, we can empathize using our imaginations, such as when we empathize by simulation. But the fact is that since robots do not experience emotions, there is no agent to empathize with, which is why feeling empathy for a sociable robot through imaginative perception does not strike me as genuine empathy. Likewise, on the modified version of Misselhorn's framework defended here, something similar applies, namely: that an illusionary empathic experience is not a case of genuine empathy. Instead, we simply misperceive that sociable robots experience emotions, are animate, and capable of humanlike mentation. Sociable robots can cause our minds "misfire," and thus, we perceive the abovementioned attributes when in fact none of them are actually present. Our misperceiving minds misperceive other minds, as it were. So although one might feel empathy or sympathy for a robot—as Frank did for his robotic companion—such cases to not strike me as genuine cases of empathy. However, when it comes to sociable robots and cognitive empathy, I must admit that perhaps a stronger case for genuine empathy could be made here. I have not explored this in very much depth in this chapter, although I think that this idea ought to become a topic of further discussion. For although robots do not really have brains, both human brains and the "silicon brains" of robots are types of informationprocessing systems. So, while robots do not experience emotions, they can arguably think, after a fashion (cf. Turing 1950 for similar discussion). Of course, perhaps the only way to settle the question of whether it is really possible to genuinely empathize with sociable robots is to keep on creating more sociable robots. For once we reach what Breazeal (2002, p. 1) calls "the pinnacle of development" of social robotics, it might turn out that robots will be able to experience emotions of their very own.

References

Bach, M. n.d. "Neon Color Spreading." Accessed October 10, 2014. http://www. michaelbach.de/ot/col-neon/index.html.

Bartneck, C., T. Kanda, H. Ishiguro, and N. Hagita. 2009. "My Robotic Doppelgänger: A Critical Look at the Uncanny Valley." The 18th IEEE International Symposium on Robot and Human Interactive Communication, 2009. RO-MAN 2009, Sept. 27 2009-Oct. 2 2009. 269-76. doi: http://dx.doi. org/10.1109/ROMAN.2009.5326351.

Bartneck, C., C. Rosalia, R. Menges, and I. Deckers. 2005. "Robot Abuse: A Limitation of the Media Equation." *Proceedings of the Interact 2005 Workshop on Agent Abuse*, Rome, Italy.

Batson, C.D. 2011. *Altruism in Humans*. New York: Oxford University Press.

Breazeal, C. 2002. *Designing Sociable Robots*. Cambridge, MA: MIT Press.

Breazeal, C. 2003a. "Emotion and Sociable Humanoid Robots." *International Journal of Human–Computer Studies* 59(1-2): 119-55. doi: http://dx.doi. org/10.1016/S1071-5819(03)00018-1.

Breazeal, C. 2003b. "Toward Sociable Robots." *Robotics and Autonomous Systems* 42(3-4): 167-75. doi: http://dx.doi.org/10.1016/S0921-8890(02)00373-1.

Burleigh, T.J., J.R. Schoenherr, and G.L. Lacroix. 2013. "Does the Uncanny Valley Exist? An Empirical Test of the Relationship between Eeriness and the Human

Likeness of Digitally Created Faces." *Computers in Human Behavior* 29: 759-71.

Darwall, S. 1998. "Empathy, Sympathy, Care." *Philosophical Studies* 89: 261-82.

Dennett, D.C. 1981. "Why You Can't Make a Computer That Feels Pain." In *Brainstorms: Philosophical Essays on Minds and Psychology*, 190-229. Cambridge, MA: MIT Press.

Dennett, D.C. 1989. *The Intentional Stance*. Cambridge, MA: MIT Press.

Festinger, L. 1957. *A Theory of Cognitive Dissonance*. Evanston: Row & Peterson.

Goldman, A.I. 1989. "Interpretation Psychologized." *Mind & Language* 4(3): 161-85. doi: http://dx.doi.org/10.1111/j.1468-0017.1989.tb00249.x.

Gray, H.M., K. Gray, and D.M. Wegner. 2007. "Dimensions of Mind Perception." *Science* 315(5812): 619. doi: http://dx.doi.org/10.1126/science.1134475.

Gray, K. and D.M. Wegner. 2012. "Feeling Robots and Human Zombies: Mind Perception and the Uncanny Valley." *Cognition* 125(1): 125-30. doi: http://dx.doi.org/10.1016/j.cognition.2012.06.007.

Iacoboni, M. 2009. "Imitation, Empathy, and Mirror Neurons." *Annual Review of Psychology* 60(1): 653-70. doi: http://dx.doi.org/10.1146/annurev.psych.60.110707.163604.

Ishiguro, H. 2006. "Android Science: Conscious and Subconscious Recognition." *Connection Science* 18(4): 319-32. doi: http://dx.doi.org/10.1080/09540090600873953.

Ishiguro, H. 2007. "Scientific Issues Concerning Androids." *The International Journal of Robotics Research* 26(1): 105-17. doi: http://dx.doi.org/10.1177/0278364907074474.

Jentsch, E.A. 2008. "On the Psychology of the Uncanny." In *Uncanny Modernity: Cultural Theories, Modern Anxieties*, translated by Roy Sellars, edited by Jo Collins and John Jervis, 216-28. New York: Palgrave MacMillan. Original edition, 1906.

Kang, M. 2011. *Sublime Dreams of Living Machines: The Automaton in the European Imagination*. Cambridge, MA: Harvard University Press.

Looser, C.E. and T. Wheatley. 2010. "The Tipping Point of Animacy: How, When, and Where We Perceive Life in a Face." *Psychological Science* 21(12): 1854-1862. doi: http://dx.doi.org/10.1177/0956797610388044.

Maibom, H.L. 2007. "The Presence of Others." *Philosophical Studies* 132(2): 161-90. doi: http://dx.doi.org/10.1007/s11098-004-0018-x.

Maibom, H.L. 2012. "The Many Faces of Empathy and Their Relation to Prosocial Action and Aggression Inhibition." *Wiley Interdisciplinary Reviews: Cognitive Science* 3(2): 253-63. doi: http://dx.doi.org/10.1002/wcs.1165.

Maibom, H.L. 2014. "Introduction: (Almost) Everything You Ever Wanted to Know About Empathy." In *Empathy and Morality*, edited by Heidi L Maibom, 1-40. Toronto: Oxford University Press. doi: http://dx.doi.org/10.1093/acprof:oso/9780199969470.003.0001.

Misselhorn, C. 2009. "Empathy with Inanimate Objects and the Uncanny Valley." *Minds and Machines* 19(3): 345-59. doi: http://dx.doi.org/10.1007/s11023-009-9158-2.

Mori, M., K.F. MacDorman, and N. Kageki. 2012. "The Uncanny Valley." *IEEE Robotics & Automation Magazine* 19(2): 98-100. Original edition, 1970. doi: http://dx.doi.org/10.1109/MRA.2012.2192811.

Redstone, J. 2013. "Beyond the Uncanny Valley: A Theory of Eeriness for Android Science Research." MA thesis, Carleton University.

Redstone, J. 2014. "Making Sense of Empathy with Social Robots." In *Sociable Robots and the Future of Social Relations: Proceedings of Robo-Philosophy 2014*, edited by Johanna Seibt, Raul Hakli and Marco Nørskov, 171-7. Amsterdam: IOS Press Ebooks. doi: http://dx.doi.org/10.3233/978-1-61499-480-0-171.

Rosenthal-von der Pütten, A.M., N.C. Krämer, L. Hoffmann, S. Sobieraj, and S.C. Eimler. 2012. "An Experimental Study on Emotional Reactions Towards a Robot." *International Journal of Social Robotics* 5(1): 17-34. doi: http://dx.doi.org/10.1007/s12369-012-0173-8.

Rosenthal-von der Pütten, A.M., F.P. Schulte, S.C. Eimler, L. Hoffmann, S. Sobieraj, S. Maderwald, N.C. Krämer, and M. Brand. 2013. "Neural Correlates of Empathy Towards Robots." *HRI 2013 Proceedings*, 8th ACM/IEEE International Conference on Human–Robot Interaction, Tokyo, Japan. 215-16. doi: http://dx.doi.org/10.1109/HRI.2013.6483578.

Saygin, A.P., T. Chaminade, H. Ishiguro, J. Driver, and C. Frith. 2011. "The Thing That Should Not Be: Predictive Coding and the Uncanny Valley in Perceiving Human and Humanoid Robot Actions." *Social Cognitive and Affective Neuroscience* 7(4): 413-22. doi: http://dx.doi.org/10.1093/scan/nsr025.

Schreier, J. 2012. *Robot & Frank*. USA: Samuel Goldwyn Films.

Slater, M., A. Antley, A. Davison, D. Swapp, C. Guger, C. Barker, N. Pistrang, and M.V. Sanchez-Vives. 2006. "A Virtual Reprise of the Stanley Milgram Obedience Experiments." *PLoS ONE* 1(1):e39. doi: http://dx.doi.org/10.1371/journal.pone.0000039.

Tinwell, A., M. Grimshaw, D.A. Nabi, and A. Williams. 2011. "Facial Expression of Emotion and Perception of the Uncanny Valley in Virtual Characters." *Computers in Human Behavior* 27(2): 741-9. doi: http://dx.doi.org/10.1016/j.chb.2010.10.018.

Turing, A.M. 1950. "Computing Machinery and Intelligence." *Mind* 59(236): 433-60. doi: http://dx.doi.org/10.1093/mind/LIX.236.433.

Yamada, Y., T. Kawabe, and K. Ihaya. 2013. "Categorization Difficulty Is Associated with Negative Evaluation in the "Uncanny Valley" Phenomenon." *Japanese Psychological Research* 55(1): 20-32. doi: http://dx.doi.org/10.1111/j.1468-5884.2012.00538.x.

Chapter 3

Robots and the Limits of Morality

Raffaele Rodogno

In this chapter, I ask whether we can coherently conceive of robots as moral agents and as moral patients. I answer both questions negatively but conditionally: for as long as robots lack certain features, they can be neither moral agents nor moral patients. These answers, of course, are not new. They have, yet, recently been the object of sustained critical attention (Coeckelbergh 2014, Gunkel 2014). The novelty of this contribution, then, resides in arriving at these precise answers by way of arguments that avoid these recent challenges. This is achieved by considering the psychological and biological bases of moral practices and arguing that the relevant differences in such bases are sufficient, for the time being, to exclude robots from adopting, both, an active and a passive moral role.

Introduction

Under what circumstances can robots properly be considered as moral agents bearing moral responsibility for their actions? Do robots have rights? Is it possible to harm or wrong a robot? I take it that most people (though, probably, robot-enthusiasts such as those reading these pages less so) will receive these questions with a certain amount of skepticism. Of course robots are *not* moral agents, and of course they *cannot* be harmed or wronged. They are after all just a bundle of circuitry and wiring typically wrapped in some poor human-looking form, if even that. There are no moral agents outside sane, adult human beings. And the only entities that can be wronged are those who have interests, such as, most clearly, sentient creatures like us. This is, of course, a crude answer. Yet it points us towards those issues that are crucial in order to answer the initial questions.

Electronic circuitry can be made to work wonders. Thanks to some such circuitry, we have built machines that beat humans at innumerable tasks. Why not allow, at least in principle, that we can build robots so sophisticated as to have whatever it takes to be part of our moral community, as agents and/or as patients? In this chapter, I shall not deny that this may in principle happen one day. I will, however, argue that that day is not today and that if that day ever comes, these robots would indeed be so similar to us that the decision to bring them about would be almost as ethically significant as the decision to bring human beings into the world.

Robots and Moral Agency

Under what conditions can anything, a child, a robot, or an adult human being, count as a moral agent? One answer to that question may go along these lines: anything that acts in accordance to what we would consider as the correct moral precepts would or should count as a moral agent. This, however, is not enough. We would not, and should not, consider as a moral agent something that just happened to act in the morally appropriate way out of sheer happenstance.

Consider next a context that offers minimal scope for moral action, such as one in which the only morally significant action is avoiding pushing a red button. Suppose you inculcate to your 2-year old child the notion that he never, under any circumstances, shall press the red button. Suppose that your child, whenever in the vicinity of the red button, diligently refrains (or avoids) pressing the red button. Once again, I hope to find my readership in agreement with the claim that we would not, and should not, consider this child a moral agent, at least not simply on the basis of his behavior around red buttons.

What is it then that makes us (or whatever else) into a moral agent? In what follows, I will not afford a complete answer to this question. The partial answer that I am about to offer, however, is arrived at by interpreting a particular structural feature of the practice called morality. I am inviting us to consider morality as a specific realm of evaluation or judgment, opposed to non-moral realms of evaluation or judgment, such as the realm of aesthetics or etiquette. What is it, if anything, that characterizes moral judgments (or the substantive moral precepts that they express) that does not characterize other non-moral judgments?

The methodology used here to answer this question offers an interpretation of what it is that we do whenever we use key moral concepts in our judgments and evaluations, and in particular whenever we use the concept 'morally wrong'. The idea is that those who, for whatever reasons, cannot grasp this central part of our practice, i.e., the evaluation of something as wrong, cannot count as moral agents, despite their capacity to act in accordance to moral precepts.

The task at hand here does not amount to a semantic analysis of 'morally wrong', a term which on many accounts is considered to be primitive and hence not susceptible of further semantic analysis. It is rather the task of stating what conditions ought to obtain in order for us to declare that an individual competently uses the concept 'morally wrong'. On this particular view, concept possession is taken to be an ability that is peculiar to cognitive agents (Brandom 1994, Dummett 1993, Millikan 2000). Possessing the concept 'cat', for example, might amount to the ability to discriminate cats from non-cats and to draw certain inferences about cats. This understanding of concepts is associated to the conception of meaning as use. To learn the meaning of a concept is to learn how to use it, which is to make justifiable assertions involving it.[1]

---

1   On this understanding of concepts, conceptual analyses are not understood as definitions stating truth conditions for the concept but rather as characterizations stating the

Consider, for example, perceptual claims. On this account, at least part of the meaning of such claims is that we are justified in asserting them when we perceive things to be in a certain way and have no defeating evidence for that appearance. On this type of analysis, justified visual experience as of red, for example, is at least partly constitutive of the concept 'red'.[2]

Going back to moral wrongness, on the view sketched here, the analysis of such concept would ultimately amount to the spelling out of the conditions under which a person would be justified in claiming that something is morally wrong. In the relevant circles, my concrete proposal is not new and is a version of what goes under the heading of neo-sentimentalism about morality. According to this view, the meaning of 'morally wrong' is (at least partly) constituted by our being justified in having certain emotions or sentiments.

Neo-sentimentalists disagree among themselves about what emotions are relevant to the analysis.[3] Hence, on one particular instantiation of this view (Gibbard 1992), to judge that Andy's hitting Ben is morally wrong is to say that it would be appropriate/fitting/rational for Ben and other impartial spectators to feel indignation at Andy and for Andy to feel guilt for her action. Others (Skorupski 2010), however, would reject that (justified) anger or indignation has a role to play in this context, while arguing that justified blame, either self-directed or other-directed, is what's ultimately constitutive of the meaning of 'morally wrong'.

Whatever the disagreements, in short, on this type of view, individuals correctly use the concept 'morally wrong'—make correct moral judgments—when they master the normative attribution of certain emotions. Hence, for example, assuming that guilt partly constituted the meaning of 'morally wrong', it would follow on this view that someone who did not understand that guilt is appropriate only for actions for which one is responsible, will not thereby master the concept 'morally wrong'. An individual such as this (a child, perhaps) may well be inclined, for example, to judge that cats are evil because they torture mice. Similarly, on this view, when facing new or complex moral questions or situations, being experienced, imaginative, and skilful in attributing the relevant emotions will help one figure out whether an action is morally permissible or not.

---

justification conditions for that concept. In order to determine the meaning of a concept, we must start by identifying the set of conditions C that must be satisfied for a person to be said to have acquired a concept. The claim is, then, that for many concepts, the conditions C are uniquely determined by the justification conditions of the concept, which can then be said uniquely to determine the concept (Pollock 1974, chap. 1).

2   This, however, is not to deny the possibility that at the level of truth conditional definitions, 'red' is semantically primitive and 'experience of red' is defined in its terms or is analytic on the meaning of 'red'.

3   Neo-sentimentalists disagree even more strongly with sentimentalists. While accepting that emotions play various roles in moral judgment, the latter deny that our moral judgments necessarily involve the normative attribution of emotions, as in "x is wrong only if it is appropriate or rational to feel emotion e".

Finally, we should note that it is *only* via the experience of certain emotions that we fully understand the meaning of 'morally wrong'. In order to show this much, let us return to the analysis of 'red' for a moment, and stress the fact that the meaning of this concept is partly constituted by justified *visual experience* as of red. Imagine a blind person who was, however, equipped with a high-tech device that communicated to her the light frequencies reflected by each of the objects she was touching with her index finger. Imagine also that this person had been taught that frequencies in the interval between 700 to 790 THz correspond to what people refer to as violet, frequencies in the interval between 600THz to 700 THz correspond to blue (and perhaps indigo), frequencies in the interval between 405 and 480 THz correspond to red, etc. For as long as this person came into tactile contact with an object, she would be able to make most of the color-related inferences that people with normal sight make. In many circumstances, then, she would appear to be a competent user of 'red'.

Yet, given the essentially visuo-perceptual nature of the concept 'red' in typical (seeing) subjects, the blind person will always miss the phenomenological aspect of 'red' and in fact will not be able to grasp the meaning of certain inferences that typical color-concepts users make based on such phenomenology. Hence, for example, we may presume that it will be a mystery for this person why or how brighter colors (which she will identify simply as those having lower frequencies) are often thought as being happy colors though also as those colors we more easily get tired of. The blind person will not be able to understand these connections between colors, on the one hand, and mental states such as happiness, tiredness, fastidiousness, or being soothing, on the other, because these connections work precisely through the specific phenomenology of the different colors.

Now an analogous point applies to the neo-sentimentalist analysis of 'morally wrong'. Consider the very idea of being wronged or the idea that an injustice has occurred. These ideas are part and parcel of the phenomenology of emotions such as anger and indignation. When undergoing emotions such as these we are not simply undergoing the evaluation that certain norms have been violated. There are all sorts of norms such as aesthetic and prudential norms, and norms of etiquette, whose violation is not experienced in terms of wrongs or injustice. But when we experience anger and indignation (and perhaps guilt and blame), we experience the relevant norm violation, as it were, in moral colors, as precisely a wrong or injustice.

To return to our analogy with colors, just as the blind person, despite her high-tech device, will fail to grasp some important ways in which we use color concepts, those lacking the relevant emotions will fail to grasp at least that part of the meaning of 'morally wrong' that is delivered to us through its phenomenology. For such individuals, it will be quite hard to distinguish violations of moral norms from other kinds of norm violations because the former will not be accompanied by their distinctive phenomenology, which include not only distinctive feelings but also specific sets of motivations (express blame, retort or punish, make amends). In other words, the emotionless will systematically miss a fundamental aspect of morality.

This point can be made more vivid through an example from literature. Think about what happens to Raskolnikov, the main character of Dostoyevsky's *Crime and Punishment* (1956), after killing the old lady and her sister in Part I of the book. The remaining six parts of the book are a vivid illustration of how it is (humanly) impossible for Raskolinkov to live with his crime. His guilt, his need to confess, and his need to atone make him at first physically ill and then almost insane. Raskolnikov's emotional life delivers and in fact constitutes the insight that he has committed a horrible deed. How can entities deprived of such emotions have access to *this* insight?

The claim being made here is that the concept 'morally wrong' and the moral practice that it animates *requires* an emotional base. This claim is crucial here, for, if it were true, robots will not be able to count as moral agents for as long as they could not be shown to feel the relevant emotions. The claim, however, can be read in two ways. On a stricter reading, each and every evaluation that a wrong or an injustice has occurred or would occur requires the occurrence of a relevant emotion (anger, indignation, or whatever emotion the neo-sentimentalist prefers). This, I think, is an implausible claim, which neo-sentimentalists do not and need not defend and which, therefore, I set aside.

On a looser reading, however, we could read the same claim as implying only that while a capacity to feel emotions is necessary to understanding moral concepts, an individual need not experience the relevant emotions on each and every occasion in which she engages in moral judgment. On this view, we would need to experience these emotions in our moral development, as we learn to grasp the peculiar nature of moral (as opposed to non-moral) judgments, and as we learn to ascribe moral wrongness correctly. Once we become acquainted with these tasks, however, while we will still *typically* experience such emotions when making moral judgments, we need not postulate their necessary occurrence. Individuals will have developed a sensitivity to these emotions, which will guide their judgments and their actions even in the absence of specific occurrences.[4] Hence, on the basis of your emotion-based experience, understanding, and sensitivity you may, at least on occasion, morally condemn someone's action quite coldly, i.e., in the absence of any occurrence of anger, indignation, guilt or blame. In short the fate of robots as moral agents hangs on their possessing the following:

Feeling: The capacity to experience the relevant emotions and their actual experiencing of such emotions during the agent's moral development.[5]

---

4 Our sense of (or sensitivity to) shame, for example, will often guide us towards certain actions and away from others. A shameless person is someone who lacks a sense of shame, and who thereby tends to act in ways that many people would find shameful (see Deonna, Rodogno, and Teroni 2011 for a discussion of this topic).

5 Coeckelbergh (2010) proposes a thesis that is on the surface similar but ultimately quite different from this. He argues in favor of the claim that robots be made to appear as if they experienced these emotions. His argument is based on the idea that we cannot be

In fact, to be more precise, in accordance with neo-sentimentalism we should say that moral agency requires both *Feeling*, and, in addition to that, the capacity to make normative attributions of the emotions at issue. As illustrated above, it is not enough to show that an individual feels guilt. He or she (or it) must also be able to judge when it would be appropriate to feel it. Hence, in addition to *Feeling*, we must also consider the following as a requirement for moral agency:

Correct Attribution: The capacity to make correct attributions of the relevant emotions.

Someone may at this point insist that moral agency requires only *Correct Attribution* and not, both, *Feeling* and *Correct Attribution*. After all, if someone were skilful at attributing the correct emotions, he or she (or it) would be skilful at making moral judgments. This line of argument should be attractive to those who want to defend the possibility that robots can become moral agents, for on the face of it, *Correct Attribution* does not seem to involve the capacity to feel emotions, which may indeed be very difficult to recreate in a machine.

There are at least two types of reply to this line of argument. First, if our capacity to make moral judgments has the phenomenological nature discussed above, then, emotionless individuals cannot quite grasp what it is involved in making moral judgments. We have defended that point by means of an analogy between moral concepts and emotionless individuals, on the one hand, and color concepts and blind individuals, on the other. If this argument were not enough, we can help ourselves to another argument from the philosophy of mind and language, namely, John Searle's famous Chinese room argument (1980).

Suppose you are alone in a room following a computer program for responding to Chinese characters slipped under the door. You understand nothing of Chinese, and yet, by following the program for manipulating symbols and numerals just as a computer does, you produce appropriate strings of Chinese characters that fool those outside into thinking there is a Chinese speaker in the room. In a similar vein, I would take it that even if emotionless robots were able to reproduce successfully the moral judgments of a human community, via their capacity to make normative attribution of the relevant emotions (if that was indeed possible in the absence of *Feeling*), they would still fail to *understand* such judgments. They would, as it were, go through the motions without grasping their meaning.

---

certain, not even in the case of human beings, that others experience emotions or, in fact, are conscious and have minds. For all we know with indubitable certainty, the moral games we play with other human beings are based as much on appearance as those we would play with robots that appeared to have the relevant emotions. In the next section, I will address this type of argument by showing that certainty is not and should not be what is at issue here. If we had good reasons to believe (rather than reasons to be certain) that an individual cannot experience the relevant emotions, but only systematically faked these emotions, we would not be able to relate to her as we would to a moral agent.

Second, we should consider why, for what purposes, we would want to endow an entity that lacks the capacity to feel certain emotions with the capacity to make correct attributions of such emotions. In human beings, the latter capacity builds on the former. For example, while children may experience guilt before age 8, they acquire those capacities necessary to make correct attributions of guilt gradually, between the age of 8 and 12 (Harris 1989, pp. 140-45, Tangney and Dearing 2002). If guilt were connected to morality in the way suggested by some neo-sentimentalists, we would come to make moral evaluations, among others, via our experiences of guilt. The correctness of such evaluations, however, depends on our capacity to learn to *feel* guilt appropriately. Suppose now that it was impossible to create robots that could feel guilt (or any other relevant emotion). What would be the point of equipping these robots with the capacity to make correct attributions of guilt? For one, these robots would not make moral evaluations via the experience of guilt.

It would make more sense to have these robots operate in the absence of any reference to emotion and operate rather with a list of proscriptions, itself devised on the basis of information concerning correct attributions of guilt in beings that *can* and *do feel* guilt. Robots operating in this way, however, will at most be accepted as agents acting within our moral community, if, that is, we trusted their capacities to recognize how to apply these precepts and how to solve the conflicts that may arise between them.[6] This is not to say, however, that they could coherently be considered to be moral agents. In the relevant respect, such robots would rather look like a version of the child in the red button example above with the difference that the robots' repertoire of proscriptions would extend beyond pressing red buttons. While such robots may be able to avoid performing morally wrong actions, we cannot infer from that that they grasp the meaning of this practice.[7]

---

6   Some caution is in order here. Many moral contexts can hardly be described as being structured by well-defined moral norms that stand in a clear hierarchical relation with each other. In such contexts agents are confronted with conflict of (moral) norms and the creative task of arriving at new norms in order to solve such conflicts. Here our capacity to feel the relevant emotions is essential in guiding us, which would in practice exclude those deprived of such emotions from taking part in such contexts, which, I assume, are not uncommon in our everyday life.

7   This conclusion may of course be challenged by those who contest the claim that emotions play any role whatsoever in determining the meaning of morally wrong. This amounts perhaps to a version of the age-old disagreement in the history of ethics between so-called sentimentalists, on the one hand, and rationalitsts, on the other. Kant's ethics is often considered to be one of the best examples of the latter tradition. It is generally difficult to solve such profound disagreements in a short space without begging the question. I will therefore refrain from attempting to do that here. It is interesting to note, however, that Kant's position may in the end not be so different from that defended above. Even according to Kant there is one moral emotion or feeling: respect (Achtung), which Kant sometimes seems to regard as the same as reverence (Ehrfurcht). Kant thinks that this emotion can motivate us. According to Sorensen (2002, p. 110), however, the moral feeling

Robots as Non-derivative Objects of Moral Consideration

Whatever one's stance about the status of robots as moral agents, the question of robots as appropriate objects of moral consideration remains open, for, after all, just as in the case of infants, to show that someone is not an active member of the moral community is not to show that he or she is not the proper object of moral consideration.[8] On the understanding of moral consideration at hand here, what is of interest is the possibility that robots be the object of *non-derivative* moral consideration.[9] In other words, showing that we ought to regulate our behavior towards robots because of the rights or interests of those who own them or are otherwise attached to them will show that the robots are the objects of moral consideration only derivatively. The same is true of arguments showing that each agent should regulate behavior towards robot because of the bad effects that failing to do so would have on the agent's character itself (Goldie 2010).

I will tackle the question of moral consideration by extending the neo-sentimentalist approach defended so far. Once again, my argument is deployed at the structural level. I will not, that is, discuss the merit of this or that particular moral norm. Rather, I will discuss the way in which some of the structural features of (this interpretation of) morality limit the content of all moral norms. To anticipate, I will show that our biology/psychology poses certain constraints on

---

plays a much more important role in Kant. Having argued elsewhere that that there is no experience without the capacity for pleasure and pain, in the third Critique, Kant argues that susceptibility or "the predisposition to the feeling for (practical) ideas, i.e. to moral feeling" is a condition for morality (Kant 1987, 5: 265). There is no morality without the capacity for the specific feeling of respect or moral feeling. "As Kant will say later, it is a misunderstanding to think anyone could have a duty to acquire these sort of feelings, since they are 'subjective conditions of receptiveness to the concept of duty' [(Kant 1996, 6: 399)]" (Sorensen 2002, p. 115). Here is, in other words, a Kantian version of the argument that emotionless robots cannot understand 'moral wrong'.

 8 Some authors (Gunkel 2014) seem to conflate the question of moral consideration with the question of (moral) rights. These, however, are slightly different questions. While being the bearer of moral rights involves being the object of moral consideration, the converse is not true: something may be the object of moral consideration without being the bearer of rights. An agent may, for example, have a moral duty to perform a certain action without a corresponding claim on behalf of any particular individual to have that duty performed. Hence, for example, charity may impose a duty to help others, without granting anyone in particular a claim or demand to be helped. (If such claim or demand existed, the action of those who addressed this claim or demand as a claim or demand owed to a particular individual would in fact cease to be considered as an instance of charity.) If this is correct, while it may be the case that robots are appropriate objects of moral consideration without being bearers of rights, the converse is not true. I therefore propose in what follows to stick to the broader question of moral consideration, so as to give robots their best chance to have moral status.

 9 By 'moral consideration' I henceforth mean 'non-derivative moral consideration', unless otherwise specified.

what can count as a moral norm. This has, in turn, repercussions on the question of moral consideration in general, and hence the status of robots as objects of moral consideration in particular.

On the neo-sentimentalist account there is no guarantee that societies will settle on an identical set of moral norms. While there are certain forms of behavior that the majority of cultures understand and have understood as immoral (e.g., murder, stealing), clearly the proscription of many types of actions (and, perhaps, thoughts and feelings) is the expression of specific cultures or cultural forms (e.g., individualistic as opposed to collectivistic cultures). This should not come as a surprise for the sentimentalist, for emotional hermeneutics are clearly influenced by cultural processes.

What is of interest to us here, however, is not so much the degree to which moral rules differ across cultures, but rather the potential degree of variety in the content of moral rules. In other words, the question is whether moral rules can have any content whatsoever merely as a function of culture or whether there are limits to what may be taken to count as a moral norm. My proposal is that while cultural processes certainly play a large role in determining the content of moral norms, they also have limits, and in particular, limits that are imposed by the biological and emotional basis of moral practices.

Consider what explanation there may be for the fact that, while most cultures morally condemn taking the life of innocents, no culture has morally condemned (as intrinsically wrong) throwing pebbles in the sea or counting to ten. This is so for at least one reason: in order to exercise moral emotive responses, we must be able to see or sense a connection between what any moral norm proscribes and some disvalue that would accrue in the absence of such a norm. In other words, we must see the norms as connected to values in some way. Hence, while we can all understand that ceasing to live is typically a disvalue, the same is not evidently true for throwing pebbles in the sea or counting to ten.

Two further points must be noted in connection with this. First, given our emotive and cognitive nature, not only is the perceived existence of a value necessary to justify a norm, but the content of the norm must also be appropriately connected to the nature of the value. Hence, if ceasing to live is the disvalue, the relative norm must forbid taking lives, or promote the preservation of life rather than, say, proscribing the game of pushpin.

Secondly, as argued above, not all norm violations are considered as the appropriate object of moral emotions. This suggests that there may well be values that dictate norms other than moral ones. Hence, for example, while many agree that beauty is a value, and there are norms to the effect that beautiful things should be admired, failure to comply with such norms is not typically considered as the appropriate object of moral emotions. It would not be appropriate for you to feel indignation at my failing to admire a beautiful thing nor would it be appropriate for me to feel guilt about it.

If this much is accepted, it follows that not all disvalue that accrues by violating norms regulating behavior is *moral* disvalue. Not all values behind these norms are significant enough or of the right kind to count as moral values. The question, then, is what kind of dis/value can legitimately count as moral? I propose that *one* category of such values are those things that matter fundamentally or are of central importance to those who are the objects of moral consideration. Let us call these things interests. When it comes to human beings, there are a variety of views as to what kinds of interests we have. Some views are monistic, as, most notably, hedonism, according to which the only interest anyone can have is the experience of pleasure and the avoidance of pain. There are, however, pluralistic views, according to which, beside pleasure and pain, we have an interest in, for example, exercising our autonomy and freedom, maintaining deep personal relations, appreciating beauty, pursuing knowledge, etc.[10] The claim I am proposing, then, is that guilt, anger, indignation, or blame will tend to be considered appropriate when a violation of norms protecting what is considered an interest of this kind would occur.

In light of this framework, the question "Who or what is the object of moral consideration?" turns into "Who or what can have interests?" If, for example, the avoidance of pain is considered an interest, then, there is a pretty straightforward argument to the effect that moral consideration should be extended to all those creatures that are susceptible of pain and similarly with regard to other interests. At this point, then, the relevant question is whether robots can be understood as having interests in any sense. This question, however, raises an important epistemological issue. How do we know what interests, if any, robots may have?

In accordance with the frame presented here, we will be inclined to answer this question by considering what is known and intelligible to us, namely, human interests. In other words, just as in the case of animals and sentience, we will ask whether robots can have any of the interests that are intelligible to us from our human perspective, whatever our understanding of this list. In what follows I will not try to show, substantively, that robots (do not) deserve moral consideration insofar as they (do not) have recognizable human interests. The point I rather want to stress is the methodological/epistemological one. When considering the question of the moral status of an entity, human beings will not have much else to appeal to than what is intelligible to them in light of their biology in its cultural declinations, namely, whether such entities have interests and what kind of humanly recognizable interests these may be. In what follows, I will try to defend this particular anthropocentric epistemology by considering and dismissing other alternatives.

---

10 Among pluralists one can further distinguish welfarists, who think that all these interests are parts or elements of human well-being, and non-welfarists, who, for example, would see well-being as a teleological value but would argue that some of these interests are non-teleological and, therefore, not proper elements of well-being. For our purposes, however, we can set these disagreements aside.

Environmental Ethics as an Alternative Framework

So far I have argued that having humanly recognizable interests grants moral consideration. I have not yet argued that interests are the *only* type of consideration that grants moral consideration. Perhaps there are other, non-interest based, types of consideration that do so. But what would these be?

In order to get some inspiration one could, for example, look at developments in environmental ethics. Research in this area is relevant here insofar as it aims at showing that nonhuman entities, such as the environment, have value and are the proper object of moral consideration. More in particular, there are positive efforts to understand the value of such entities independently of human interests (Rodogno 2010). The environment, that is, would be good even if it did not contribute to our well-being in any way, and, in fact even if no human being were there to value it.

Claims such as this, however, are not of the kind that we are looking for, as they may well be compatible with the idea that the environment is valuable insofar as *it* has interests of some kind. Now this kind of thinking is not an alternative to the one offered above. Rather, it confirms precisely the idea that our moral thinking is forced into adopting such categories. Perhaps an alternative way of thinking is offered by a holistic approach to the value of the environment such as Callicott's (1989, p. 25), according to which the *summum bonum* is the integrity, stability, and beauty of the biotic community. What ultimately matters on these views are the integrity, stability, and beauty of the whole, not the well-being or interests of individuals. How we are normatively to relate to the individuals (and their interests) composing that whole is regulated by the valuable features of the whole.

Such views raise an immediate question: Why is the stability and integrity of anything a value? Why not choose instability and disunity? And why think that these are *moral* values (the fact that Callicott includes beauty as part of the trio is not promising)? My guess is that if we tend to find at least some initial appeal in the thought that stability and integrity are values, it may be because we believe that a certain degree of stability is necessary to the pursuit of our interests and then extend this framework to the environment. This, however, is no alternative type of thinking.

Even granting some initial appeal to the idea that states of systemic stability and integrity are valuable, another question remains. How do we, as moral agents, relate to such values? What norms follow from such values? Remember how we argued that moral norms must regulate behavior in a way that is appropriate to the values they protect. What are we to do in order to respect the stability and integrity of ecosystems? The quick answer is of course that we have reason to *preserve* them. One way to read this claim is to take it as meaning that we should preserve them as they are now. But ecosystems are not unchanging systems. All of them were not in existence at some point in time, they mutate over time, and go out of existence at some point in time, even without any human interference. So, preservation of their *status quo* is not necessarily the right attitude.

Perhaps the claim should be read as meaning that we should see to it that we *not destroy* them. Rather than preservation, then, we should say non-destruction. The destruction or disappearance of ecosystems, however, is a fact of nature, independently of human activity. Intuitively, then, non-destruction cannot mean that we should actively do whatever it takes to extend the life of an ecosystem, just as valuing human well-being and life does not necessarily mean we have to extend human life *ad infinitum*.

Perhaps, then, the idea of *non-interference* is the best (deontic) interpretation of the claim that ecosystems are ultimate values. But the idea that we ought to refrain from interfering with ecosystems would be a strange one in many respects. First, by being on the same planet, we are bound to interfere with at least some ecosystems. We are part of this world and its nature as much as any other terrestrial being. Now, and this is a second reason against non-interference, all sorts of animals not only interfere but are constitutive elements of ecosystems. Their lack of an active role may cause the ecosystem to collapse. Third, human interference with an ecosystem is not necessarily detrimental. Some grasslands have been sustainably exploited for thousands of years (Mongolia, Africa, European peat and moorland communities).

Returning to our discussion of the moral status of robots, we should take this excursus into environmental ethics as an illustration of the kind of problems that are likely to arise if we tried to abandon the framework defended so far. First and foremost it seems indeed that our moral thinking is bound to the category of interests. If we try to unshackle it from this category, a number of difficult questions arise. If robots do not have interests, on the basis of what should they be the object of moral norms and consideration? Supposing that they did have non-interest based values, what would they be? And how could we determine the content of the norms that such values would dictate to us? Finally, would these norms be rightly understood as moral as opposed to non-moral norms?

A Final Hurdle

Before these difficult questions, one may conclude that the safer alternative is to accept the limits of our moral thinking and stick to the type of frame defended above. This, however, would be a hasty conclusion, for there may still be an alternative approach and, in the absence of that, we may want to reject the frame defended here on other grounds. In what follows, I will therefore present, in turn, some recent criticisms that have been put to an approach which, at first sight, seems to be very close to ours; and, then, present and reject a last alternative approach. The Standard Approach is characterized as follows:

> entity x has property p
> any entity that has property p, has moral status s
> entity x has moral status s. (Coeckelbergh 2014, p. 63)

On the face of it, our approach may be seen as an instance of the Standard Approach, for, on our approach, we would indeed claim that any entity that has the property of having interests, has moral status, and, even more strongly, that only those entities that have that property have moral status. We would then go on to examine whether robots display such properties and, for as long as we have reason to believe that they don't, we would deny that they are non-derivative objects of moral concern.

Both Gunkel (2014, pp. 120-21) and Coeckelbergh (2014, p. 63) believe that this approach falls prey to serious epistemological problems affecting the first and second steps. In particular both authors espouse a sort of skepticism that implies that we cannot be certain and cannot have indubitable knowledge of other minds, and in particular or whether other minds have the capacity to suffer or are otherwise conscious. This obviously affects the first step of the standard view, as many of the interests that we recognize are indeed dependent on sentience and/or on having a mind. The same kind of skepticism, however, affects also the second step. As Coeckelbergh puts it:

> How do we know for sure that a particular property p justifies moral status s? Do we have access to a moral metaphysics, a Book of Values, in which we can find propositions about moral status that cannot be doubted? And how can we be so sure in the case of new entities such as autonomous intelligent robots? Again, a skeptic response seems in order here. (ibid.)

Our approach, however, can eschew these epistemological problems. Sure enough, we cannot, *with certainty*, make claims about sentience and other minds but we do not need to make such indubitable claims. On our approach it is sufficient to have justified beliefs about sentience and other minds. Inference to the best explanation would be enough here to justify our believing that other human beings and animals sufficiently similar to us have mental states. The argument would go as follows:

> Other human beings are very like me. They behave very much as I do in similar circumstances *and they are made of the same stuff*. When I burn myself it hurts and I cry out and wince. When other people are burned they do the same. I can thus infer that they are in pain too. There are multifarious such similarities. Put more generally, I know directly that I have beliefs, emotions, feelings, sensations and the like. So I am enabled to infer, on the basis of these multifarious similarities, that other people also have beliefs, emotions, experiences and the like. In short, I am entitled to infer that other human beings have as I do, an inner life and that it is very like mine. (Hyslop 2011, sec. 1, emphasis added)

Now, for as long as robots are not sufficiently similar to us in the relevant respects, i.e., made of the same stuff, we have no reason to believe (rather than indubitable knowledge) that they have minds and/or are sentient and, hence, cannot attribute to them the kinds of interests that give rise to moral status. Similarly, the claim that we, as human beings, have no real alternative other than conceptualizing moral status in terms of humanly recognizable interests is *not* an *a priori*, indubitable intuition read off a Book of Values, but an *a posteriori* critical interpretation of our moral thinking, open to challenge by other interpretations.

The troubles for the Standard Approach and its affiliates, may, however, not be over. On this type of view, there is an alleged unbridgeable *explanatory gap* between the attitudes of people that attribute mental states and engage affectively with robots, on the one hand, and the idea that robots, given their lack of minds, sentience, and interests, are "mere machines." When it comes to robots and their moral status there is, in other words, a gap between what certain people experience when engaging with them and our reasoning about them, between thinking and action, between belief and feeling. We may *think* about them as mere machines, when thinking about them in a scientific mode, and yet experience particular robots as "more than machines" and address some of them with "he", "she", or even "you." Coeckelbergh goes on to write:

> How should we respond to this gap between beliefs and behavior, between reasoning and experience? The moral–scientific answer to this problem is then that we are simply incorrect about the entity's status. There is a gap, so the answer goes, but there should not be a gap. But this answer does not help us to understand how we act towards these robots and it makes all responses except the 'correct' one appear as irrational. We cannot make sense of experiences that depart from the idea that a robot is a machine and we have to dismiss them as 'childish' or 'ignorant'. We have to say: 'Don't you know this is a machine?' But is this the only possible answer? Is it the best answer?" (2014, p. 64)

This explanatory gap, however, is not a necessary feature of the Standard Approach, not, at least, in the version defended here. As argued elsewhere (Rodogno forthcoming), there are a variety of ways in which we can conceptualize human-to-robot affective engagement that fall short of attributing irrationality or immaturity to the human. Hence, for example, we could hypothesize that, when engaging affectively with robot pets, individuals adopt a cognitive mode akin to that which is normally adopted in our engagement with fiction. Being emotionally engaged by robot pets would be akin to being emotionally engaged by a good novel or movie. Just as my sadness for Anna Karenina involves my *imagining, accepting, mentally representing* or *entertaining the thought, without believing*, that certain unfortunate events have occurred to her, my joy at the robot pet involves my imagining, accepting, mentally representing or entertaining the thought, without believing, that it is happy to see me. No need to misrepresent the world or be irrational about anything.

I suspect, however, that this type of answer may still fail to satisfy the critics of the Standard Approach. According to Coeckelbergh, philosophers adhering to this approach simply adopt the wrong kind of methodology, as they proceed by isolating robots from their contexts and situatedness, in the same manner in which scientists isolate the object of their study from its environment by creating artificial, experimental conditions. By proceeding in this way, philosophers lose sight of precisely the fact that the object of their study stands in specific relations with its environment. An elderly attached to a robotic pet, who treats the "pet" as a real pet or baby, for example, is already engaged in a relation with the robot that goes beyond the question of the moral status of the robot. At this point Coeckelbergh suggests:

> Is not moral quality already implied in the very relation that has emerged here? … What needs to be discussed is that relation, rather than the moral standing of the robot. (2014, p. 70)

On the alternative approach Coeckelbergh proposes:

> [… T]here are several ways in which an entity can appear to us, with none of these ways of seeing having a priori ontological or hermeneutical priority. Some ways of seeing may be better than others, but this evaluation has to take place with regard to particular entities, practices and experiential situations, can allow of various perspectives, and cannot be pre-*determined* "before" by a metaphysical properties ontology. (ibid., p. 65)

On this alternative approach there is no "correct" way of seeing the robot. Moral status ascriptions are at one level growing within the relation. This, I believe, is the point at which the Standard Approach and the approach defended here will stop accommodating the criticisms made by Coeckelbergh. Why think that the fact that an individual is attached to a robotic pet in the same way as he would be attached to a real pet or a baby makes the question of the moral standing of the robot irrelevant or superfluous? Is it because it is already so clear that for the individual who is attached to the robot the latter already has moral status? But is that correct? If the suggestion made above is descriptively accurate, it may well be that the individual in question engages with the robot in the same way in which we engage with fiction. Her emotions would be real but nonetheless resting on cognitive bases that are not regulated by a truth-norm. The fact that the robot pet actually has no interests and hence no moral status would not give her any reason to change her attitudes towards it or be less attached to it.

Even admitting that those attached to robot pets took the latter to have moral status, the question of the moral status of the robot would still be entirely pertinent. For one, in line with the view discussed above, we should understand what moral norms would follow from the fact that this particular individual is attached to this particular robot. Does it, for example, follow that you and I and everybody else now have reason to regulate our behavior towards the robot? The answer, I would imagine, is positive: we should regulate at least some of our behavior with regard to the robot. But in order to understand in what respects we should do so, we would need to understand whether the robot is a derivative or non-derivative object of moral consideration, for the type of norms that would follow in either case would indeed be different. Now in order to argue that the robot has non-derivative moral status, it would not be enough to show that its relation with one individual already involves "moral quality".

Morality is first and foremost a practical affair, a system of rules whose rationale is the regulation of behavior. It works through the establishment of norms that will at times restrict the agents' freedom to pursue their own interests. A stance according to which there is no "correct" way of seeing the robot and its moral status, which makes appeals to a supposed "moral quality", will fail to be sufficiently practical. As argued above, agents will fail to understand and, consequently, be moved by such appeals unless they could see how robots have recognizable interests. Yet they will, for the same reason, understand and be moved by claims to the effect that we should "respect" the robot for the sake of the person who is attached to it.

For as long as we have no reason to believe that robots have interests or other plausible alternative ways of thinking are presented, we shall rest content with the idea that we should regulate our behavior towards robots only derivatively, that is, for reasons having to do with the interests of those who are non-derivative objects of moral consideration.

Conclusion

In conclusion, then, our biology and psychology pose limits to human morality in two important respects: by excluding from the realm of non-derivative objects of moral attention anything incapable of humanly recognizable interests, and by excluding from the realm of moral agents anything incapable of feeling a certain range of emotions. As argued at the start, this is not to say that robots will always be excluded from our moral community in these two important respects. That question is not one I have examined here. The day in which robots fulfill the conditions for moral agency and moral patience outlined here, however, will be the day in which they will be emotive creatures capable of humanly recognizable interests. These robots would indeed be so similar to us that the decision to bring them about would be almost as ethically significant as the decision to bring human beings into the world.

References

Brandom, R.B. 1994. *Making It Explicit: Reasoning, Representing, and Discursive Commitment*. Cambridge, MA: Harvard University Press.

Callicott, J.B. 1989. *In Defense of the Land Ethic: Essays in Environmental Philosophy.* . Albany, NY: SUNY Press.

Coeckelbergh, M. 2010. "Moral Appearances: Emotions, Robots, and Human Morality." *Ethics and Information Technology* 12(3): 235-41. doi: http://dx.doi.org/10.1007/s10676-010-9221-y.

Coeckelbergh, M. 2014. "The Moral Standing of Machines: Towards a Relational and Non-Cartesian Moral Hermeneutics." *Philosophy & Technology* 27(1): 61-77. doi: http://dx.doi.org/10.1007/s13347-013-0133-8.

Deonna, J., R. Rodogno, and F. Teroni. 2011. *In Defense of Shame: The Faces of an Emotion.* NY: Oxford University Press.

Dostoyevski, F. 1956. *Crime and Punishment.* Translated by Constance Garnett. New York: Random House. Original edition, 1866.

Dummett, M. 1993. *Seas of Language.* Oxford: Oxford University Press.

Gibbard, A. 1992. *Wise Choice, Apt Feelings: A Theory of Normative Judgment.* Cambridge, MA: Harvard University Press.

Goldie, P. 2010. "The Moral Risks of Risky Technologies." In *Emotions and Risky Technologies*, edited by Sabine Roeser, 127-38. Dordrecht: Springer Netherlands. doi: http://dx.doi.org/10.1007/978-90-481-8647-1_8.

Gunkel, D.J. 2014. "A Vindication of the Rights of Machines." *Philosophy & Technology* 27(1): 113-32. doi: http://dx.doi.org/10.1007/s13347-013-0121-z.

Harris, P.L. 1989. *Children and Emotion: The Development of Psychological Understanding.* Oxford: Blackwell.

Hyslop, A. 2011. "Other Minds." In *The Stanford Encyclopedia of Philosophy*, ed Edward N. Zalta. Spring 2014 Edition. http://plato.stanford.edu/archives/spr2014/entries/other-minds.

Kant, I. 1987. *Critique of Judgment.* Translated by Werner Pluhar. Indianapolis: Hackett. Original edition, 1790.

Kant, I. 1996. "Metaphysics of Morals." In *Practical Philosophy*, translated by Mary J. Gregor, 353-603. Cambridge: Cambridge University Press. Original edition, 1797.

Millikan, R.G. 2000. *On Clear and Confused Ideas: An Essay About Substance Concepts.* Cambridge: Cambridge University Press.

Pollock, J.L. 1974. *Knowledge and Justification.* Princeton, NJ: Princeton University Press.

Rodogno, R. 2010. "Sentientism, Wellbeing, and Environmentalism." *Journal of Applied Philosophy* 27(1): 84-99. doi: http://dx.doi.org/10.1111/j.1468-5930.2009.00475.x.

Rodogno, R. forthcoming. "Social Robots, Fiction, and Sentimentality." *Ethics and Information Tecnology.*

Searle, J.R. 1980. "Minds, Brains, and Programs." *The Behavioral and Brain Sciences* 3(3): 417-57. doi: http://dx.doi.org/10.1017/S0140525X00005756.

Skorupski, J. 2010. *The Domain of Reasons.* Oxford: Oxford University Press.

Sorensen, K.D. 2002. "Kant's Taxonomy of the Emotions." *Kantian Review* 6: 109-28. doi: http://dx.doi.org/10.1017/S136941540000162X.

Tangney, J.P. and R.L. Dearing. 2002. *Shame and Guilt.* New York: The Guilford Press.

Chapter 4

What's Love Got to Do with It? Robots, Sexuality, and the Arts of Being Human

Charles M. Ess[1]

I explore contemporary AI and social robots in order to clarify what we may expect of these in terms of love and friendship. I first review Anne Gerdes' discussion of *phronesis* and reflective judgment as remaining computationally intractable (2014). *Phronesis* further implicates virtue ethics; these likewise work in John Sullins' arguments that human *eros* cannot be instantiated in social robots (2012, 2014). I complement these two threads with Sara Ruddick's phenomenological account, as foregrounding embodiment, autonomy, and self-awareness as necessary conditions of "complete sex"—including the specific desire that our desire be desired by the Other (1975). Complete sex then implicates mutuality, equality, respect for persons, and loving as virtues. I connect these in turn with patience, perseverance, and empathy as virtues requisite for human friendship (Vallor 2009, 2011a, b, 2015). Social robots arguably lack all of these capacities—including autonomy and embodied awareness of desire—and so cannot qualify for complex sex and human *eros*. This analysis thus demarcates human and machine capabilities more fully, and underscores the recognition that if we are to remain distinct from sexbots, we are required to *become* better friends and lovers through the cultivation of the virtues requisite thereto.

Introduction

In the traditions of the computational turn, as illumined by recent developments in cognitive science and roboethics, I mark out what appear to remain as intractable differences between human capacities and abilities and those we can instantiate and replicate in Artificial Intelligence (AI) and social robotics. The overarching goal is to thus use social robots and AI as empirical test-beds, against which we can test our best intuitions and sensibilities regarding what it means to be human.

To begin with, I draw on the work of Anne Gerdes (2014) to highlight *phronesis*—the distinctive capacity of reflective ethical judgment—as a primary

1   I am very grateful indeed to Anne Gerdes, John Sullins, and Marco Nørskov for their critical comments and suggestions in response to an earlier version. The chapter is considerably clearer and more substantive as a result.

point of demarcation between human and machine: in turn, John Sullins terms AI-based versions of reflective judgment as "artificial phronesis" (2014, p. 7). Sullins further conjoins artificial phronesis with artificial autonomy and artificial agency as terms that suggest a weak continuity between human and machine.

I then take up matters of love and sexuality, beginning with Sullins' arguments for human *eros* as remaining distinct from the sorts of sexual engagements we might enjoy with social robots (Sullins 2012). A critical component of Sullins' account is its appeal to virtue ethics: I seek to enhance and compliment Sullins' arguments by turning to both virtue ethics and phenomenology (consistent with neo-phenomenological approaches in cognitive science). Specifically, Sara Ruddick's phenomenological analysis of "complete sex" contrasts more dualistic understandings of selfhood and thus sexuality with more non-dual experiences (1975)—experiences captured by phrases such as "I *am* my body" and phenomenologist Barbara Becker's neologism *LeibSubjekt*, the "BodySubject" whose personhood infuses his or her distinctive body, including its sexuality (2001). Complete sex takes place within this latter mode and is thus marked by the full presence and engagement of persons as autonomous, self-aware, emotive, embodied, and unique. Most importantly, Ruddick highlights the central role of mutual desire in complete sex between two such fully present persons: we not only desire the Other—we desire to be desired and, still more completely, we desire that our desire be desired. On Ruddick's showing, this mutuality of desire and desirability in complete sex entails the critical *virtues* of respect for persons, equality, and loving. I link these virtues with those argued by Shannon Vallor (2009, 2011b) to be necessary for deep friendship, namely, empathy, patience, and perseverance.

On this showing, social robots may well serve as satisfying partners for "just sex" or good sex: but they fail to meet the criteria of complete sex as they lack full autonomy and full self-awareness, including awareness of genuine emotions, and most specifically of desire. These deficits in turn render social robots incapable of serving as Others who call us into erotic relationships that foster the virtues of mutuality, respect, and loving, as well as those of empathy, patience, and perseverance. Indeed, rather than fostering these critical virtues, sexbots by themselves risk short-circuiting our development in these directions and thereby diminishing our capacities for good lives of flourishing as both friends and lovers.

This analysis thus offers a still more complete account of both similarities and critical differences between human and machine—so as to highlight the roles of mutual desire in complete sex as coupled with virtues requisite for love and friendship in defining our humanity as distinct from robots, however capable, sophisticated, and beneficent the latter may be on other grounds and in other contexts. At the same time, however, to establish and sustain these differences sets up very high ethical bars—beginning with the recognition that being friends, lovers, and humans are not givens, but demanding and difficult projects of acquiring and fostering specific virtues. But this means, finally, that our exploration of what appear to remain intractable differences between humans and machines may not only help us better understand these difference and thereby more fully and precisely foreground our distinctively human capacities and abilities: in addition, especially as these emerge as *virtues*—practices and habits that can be pursued and cultivated—these understandings may inspire us in turn to become better human friends and lovers.

Backgrounds

The arguments I present here draw from three large backgrounds. The first I consider to be a classic "computing and philosophy" (CAP) approach that emerged under the rubric of "the computational turn" in philosophy by the late 1970s and early 1980s. This approach was inspired first of all by the rise of digital computers themselves as logic machines in general, and by the emergence of Artificial Intelligence (AI) as a critical intersection between philosophy, psychology, computer science, and related disciplines at least as far back as the 1950s. With some lag, philosophers increasingly turned their attention to the array of critical thematics and issues evoked by computers both specifically and more broadly, e.g., questions of machine logic, the nature of mind, the ethical challenges and social implications of infusing human societies with increasingly sophisticated computational devices, etc.: these eventually consolidated in what we came to refer to as the computational turn in philosophy by the late 1970s and early 1980s— a turn explored with particular focus in the first "Computing and Philosophy" (CAP) conferences.[2] To be sure, developments and advancements in all of these fields—most especially computational technologies and robotics—have only increased in their pace over the past decade or so. But while these developments helpfully force us to continuously reflect and revise earlier arguments and findings—the questions explored here remain more or less the same and simple: generally formulated, how far—and how far not—are we able to instantiate in computational technologies what we think of as specifically *human* capabilities and components, beginning with "intelligence," however construed?

This classic strategy thus seeks to better understand what it means to be human by examining how far—and how far not—we may instantiate what we take to be important human abilities in machines. To paraphrase the watershed volumes by

---

2 It may be helpful to note here that CAP expanded and thrived as did the ongoing developments and interactions between computing and philosophy as well. CAP eventually became the International Association for Computing and Philosophy (IACAP) and was joined by two sister organizations devoted to exploring especially the ethical dimensions of computing—INSEIT, etc., and ETHICOMP, etc. I have participated in most of these conferences in a number of ways, from presenter to organizer to editor of subsequent publications, etc. Much of the broad background that shapes the approaches in this chapter has been developed over the nearly 30 years' history of IACAP and the 20 years' history of CEPE and ETHICOMP.

Hubert Dreyfus (1972, 1992), to ask the question "What can computers (still) not do?" is to take up our best efforts to replicate (if not enhance) human capabilities as a kind of test-bed within which we can explore in ways deeply informed by *empirical* findings and accomplishments what may indeed be comparable across different platforms of instantiation, and what may remain (for the time being at least) distinctively human. Wendell Wallach (2014) captures this approach nicely: "Research on AI, and in particular on humanoid robots, forces us to think deeply about the ways in which we are similar to and truly differ from the artificial entities we will create."[3]

More specifically—and this brings us to the second and third backgrounds—many of my earlier intuitions and arguments regarding what computers, and, by extension, robots, can and cannot do were grounded in (2) understandings of computation that were deeply rooted in Turing's formal definition of computation, "as the step-by-step symbol manipulation carried out by a Turing machine, conceptualized in abstract mathematical terms" (Boden 2006, p. 1414) and (3) as the work of Hubert Dreyfus exemplifies, especially phenomenologically-based approaches to articulating our being human. As Boden demonstrates, however, there have been important shifts on both grounds—shifts that, most briefly, have helped us move from a more confrontational stance between what is now called "good old-fashioned AI" (GOFAI) and phenomenological analyses towards more constructive and complimentary approaches. Most briefly, our understanding of "computation," while still dependent on Turing for a formal definition, now extends to forms of computation that are not easily reducible to a Turing definition: Boden discusses hypercomputers and quantum computers as examples (2006, p. 1417f.). These foundational transformations, as Boden says, mean that 'computation' is "a moving target" (2006, pp. 1414-28). At the same time, there has emerged what Boden calls "the neo-phenomenological movement in the philosophy of cognitive science," including John Haugeland, Timothy van Gelder, and Andy Clark (ibid., pp. 1399ff.). Especially important for my purposes here is the emergence within this movement of notions of *embodied cognition* (e.g. Wilson and Foglia 2011) and the *embodied* mind—specifically, Andy Clark's account of the *extended mind* and its supporting "computational philosophy of embodiment" (Boden 2006, pp. 1404-7). While the details of this approach are, of course, contested—the account stands as an important *conjunction* of neo-phenomenological approaches that draw

---

3   See as well Coeckelbergh (2012). Specifically, this focal question defined the conference Robo-Philosophy 2014—Sociable Robots and the Future of Social Relations (August 20-23, Aarhus University, Denmark). So one of the conference organizers, Johanna Seibt, defined the specific meaning of robo-philosophy as "the response to the realization that the possibility of artificial social agency forces us to question the ascription conditions of moral status, cognitive capacities, sociality, normative agency, and responsibility at the most fundamental, metaphilosophical level. The term 'robo-philosophy' is to invite reflections about a possible turning point in the discipline of philosophy itself" (2014, p. viii).

on the work of Heidegger, Merleau-Ponty, and Dreyfus, and rigorous approaches to computation, further enriched by Clark's work in neuroscience and biology.

On the one hand, I take this shift from more confrontational to more synthetic or complementary approaches as critical background as it is more open to especially the sort of phenomenological exploration I pursue below as a way of examining our being embodied humans *vis-à-vis* what computers and robots may or may not be capable of achieving (yet). At the same, however, these shifts by no means imply that even the most robust computational account of mind or intelligence will necessarily lead to a fully human-like artificial mind. As Boden concludes her magisterial overview of diverse contemporary views on these matters, she highlights Peter Godfrey-Smith's taxonomy of approaches to A-life (artificial life) (1994), including "weak and strong versions of the continuity of life and mind" (Boden 2006, p. 1442): " ... the weak continuity theorist sees mind as emerging only from life, but as significantly different from it, whereas the strong continuity theorist regards mind and life as ontologically similar, sharing basic organizational principles" (ibid.). For example, a Heideggerian insistence that *Dasein* is uniquely human, and/or a Wittgensteinian insistence that intentionality is only possible for humans capable of discerning and constructing linguistic concepts and meaning would thus work as components of weak continuity theories. For her part, Boden appears to lean in the direction of a weak continuity theorist, as she aptly summarizes:

> In sum, the relation between life and mind is still highly problematic. That applies to work in AI/A-Life, and to philosophy too. The common-sense view is that the one (*life*) is a precondition of the other (*mind*). But there's no generally accepted way of proving that to be so. (2006, p. 1443, cf. Searle 2014)

In this light, then, it remains both possible and necessary to explore how far important human capacities and experiences may further illuminate our understanding of being human *vis-à-vis* what our computers and machines can and cannot do.

What can Computers and Robots (Still Not) Do?

I first turn to matters of ethical judgment and *phronesis* as strong candidates, along with analogical reasoning, of what cannot be fully instantiated in AI and thus social robots. I then take up matters of love, sex, and *eros* from the perspectives of both virtue ethics and phenomenology, as these highlight the requirements of mutual desire between embodied persons and correlative ethical moments of respect for persons, equality, and loving (especially in the sense of *eros*) as a virtue. We will see that robot lovers lack the necessary conditions for an Otherness that includes awareness of desire and thus for mutuality of desire; they are thus incapable of fostering the virtues of respect for persons, equality, and loving. On the contrary, robot lovers might well short-circuit the pursuit of these virtues and thus dilute or reduce our own development as virtuous friends and lovers in good lives of flourishing.

To begin with, Anne Gerdes has recently argued that despite the remarkable advancements in robotics and AI, most especially in conjunction with so-called "Lethal Autonomous Robots" (LARs: or LAWs—Lethal Autonomous Robots) designed for warfare, a number of important blocks remain between what is computable and what we require for human ethical *judgment* and responsibility.[4] To begin with, especially when invoking such traditions and notions as Just War Theory, it is painfully clear that a simple "top-down" model for computing when—and when not—lethal force may be legitimately used is inadequate. Such top-down approaches may well work with making *determinative* judgments—judgments that generally proceed in deductive fashion from general principles to specific conclusions. But as I have discussed elsewhere (Ess 2013, pp. 28-30, 239), *phronesis*, as a specific form of *reflective* judgment, rather works "bottom-up." That is, such reflective judgment is tricky and difficult—and, apparently, not reducible to deductive or algorithmic approaches—as it begins with the fine-grained details of a specific situation and context: from there, we as humans then use *phronesis* to discern what more general principles, norms, and rules may apply. But this in turn is not sufficient to draw a conclusion or make a judgment: rather, we must further seek to discern among a particular collocation of general principles, norms, and rules which of these should hold greater weight over the others. This *judgment* is profoundly dependent upon the specific details of a current situation: and so, again, there appears to be no "über-algorithm" or determinative/deductive procedure for defining with unambiguous certainty just which norm(s), principle(s), and/or rule(s) should take priority in our given case.

These difficulties are further compounded by three additional features of reflective ethical judgment. First, as Gerdes notes with regard to Just War Theory requirements for *proportional* use of force (i.e., no more than is absolutely necessary) and *discrimination* between combatants (as legitimate targets) and non-

---

4   This focus on human judgment as (likely) not reducible to computational techniques is part of Joseph Weizenbaum's arguments in his seminal volume, as announced in the title, Computer Power and Human Reason: From Judgment to Calculation (1976). Weizenbaum roots his emphasis on judgment in Hannah Arendt's observations of policy makers in the Pentagon as eager to embrace a quasi-mathematical understanding of reason and human political and historical realities as reducible to laws as mathematically expressible and necessary as those of the natural sciences: Arendt further notes "[They] did not judge; they calculated … . An utterly irrational confidence in the calculability of reality [became] the leitmotif of the decision making" (Arendt 1972, pp. 11ff., cited in Weizenbaum 1976, pp. 13f.). Weizenbaum contrasts human judgment explicitly with what is calculable (1976, p. 44)—but does not further elaborate on how human judgment may resist reduction to the calculable and computational. In this light, the attention to judgment and phronesis at work in this chapter may be seen as an extension of Weizenbaum's and Arendt's original observations and intuitions.

combatants (as non-legitimate targets)—"there is no clear yardstick in place for measuring how to follow the rules of proportionality and discrimination" (2014, p. 284). Rather, these are often tricky (and sometimes fatally wrong) judgments that depend, as Gerdes points out, on both "situational awareness and experience based knowledge, i.e., *phronesis*, to make these kinds of decisions" (ibid.). This is to say: making use of *phronesis* requires both (long) experience and acute awareness of the immediate context: but manifestly, the experience of one human being will vary from the experience of another—and the upshot is that reflective or phronetic judgments can be legitimately variable, even within the same context. This further entails, however, that each individual is thereby uniquely *responsible* for his/her/its phronetic judgment, as it flows from and is dependent on his/her/ its lived experiences. This is somewhat tricky in the case of LARs, insofar as it is further unclear just how far fully-fledged *autonomy* can be achieved by or granted to an AI (a point we will return to shortly): nonetheless, as Gerdes notes "a notion of free choice is still called for in order to represent responsibility properly" (ibid.).

Third, Gerdes cites Drew McDermott's argument that ethical reasoning is distinctive from other forms of reasoning—specifically, computational reasoning—insofar as it "involves thorny problems such as analogical reasoning, and deciding the applicability of imprecise precepts and resolving conflicts among them" (McDermott 2008, in Gerdes 2014, p. 284). As I understand it, in ethics we employ analogical reasoning both broadly in the exercise of casuistics, the effort to draw ethical conclusions in new contexts based on (hopefully) close analogies with previous cases that serve as precedents—and specifically in *phronetic* or reflective judgments: it would seem that if we are to resolve an ethical dilemma or challenge through such reflection, a central move here is our effort to determine if there are other cases or examples in our experience that the current situation and challenge are *like* in a strongly analogical way. Insofar as we *judge* that likeness holds, we can further judge what the successful resolution of an earlier challenge would look like (i.e., its analogue) in our current case.

The history of philosophical efforts to come to grips with analogical reasoning is as old as the Pythagoreans and then Plato and Aristotle: it extends through such thinkers as Aquinas and Kant who seek to use especially *qualitative* analogies—in contrast with their mathematical *quantitative* counterparts—to resolve an array of difficult problems (cf. Ess 1983). What is especially important for our purposes is that fully instantiating analogical reasoning in computational systems remains a very knotty problem.

To be sure, as Gerdes notes, so-called "bottom-up" (e.g., connectionist) approaches to AI appear to offer important advantages over more top-down approaches—which has led Wendell Wallach and Colin Allen, for example, to propose a hybrid model for AIs (2009). As Gerdes remarks, Wallach and Allen take up their hybrid model inspired in part precisely by Aristotelian virtue ethics— i.e., an ethics which highlights precisely the role of *phronesis* as a reflective form of judgment. It is somewhat curious, however, that Wallach and Allen do not explicitly address the matters of reflective judgment or *phronesis* in their treatment. Whatever the reasons for this may be, I share with Gerdes here her more general observation that, given these diverse challenges, " ... one can have serious doubt that *phronesis* is at all computationally tractable in any architecture" (2014, p. 284).

Indeed, *phronesis* and its allied facilities and roles are taken up by other scholars and researchers as one of several points of limitation between human and machine (e.g. Kavathatzopoulos and Asai 2013, Kavathatzopoulos 2014). In particular, John Sullins—partly in response to my earlier (Ess 2009, p. 25)—has pursued a virtue ethics approach to develop morality especially for social robots.[5] Using the term from Godfrey-Smith, we can understand Sullins' view as one version of a weak continuity theory. That is, Sullins argues that we may not achieve fully human-like capacities in AIs and social robots: but we can come close enough in important areas, beginning with what he calls a "functional free will" (2014, p. 7). This free will is not, as he puts it, "a hard-core Sartrean free will," but one nonetheless sufficient for grounding an "artificial ethical agency": this agency is "semiautonomous" in the sense that it is programmed for a specific task in a controlled environment—thereby avoiding precisely the problems Gerdes details for LARs operating in uncontrolled, real-world environments and attempting to pursue a range of tasks. In parallel, such a social robot would enjoy "artificial *phronesis*"—i.e., a limited form of phronesis as compared to its human counterpart, but one sufficient, in Sullins' view, for "artificial moral agency" to such a robot (Sullins 2014, p. 7). Completing the parallelisms between a virtue ethics view of human beings and a virtue ethics approach to developing artificial ethical morality and agency, Sullins suggests that such a social robot would begin with *programmed* virtues, such as security, integrity, accessibility, and ethical trust. But in the long run: just as human beings learn what important virtues are and must practice acquiring them in relationship with other human beings, including the *phronemos*, the exemplary person who best exemplifies and embodies at least some of these virtues—so our social robots would seem to require their own *phronemos* as well. Indeed, it will be interesting to see how far we might decide that the *phronemos* should be a virtuous human—and/or a virtuous social robot, i.e., one who has learned both familiar virtues from the human world and, perhaps,

---

5   As with the attention to judgment and phronesis, invoking virtue ethics in these contexts is neither idiosyncratic nor without important precedent (cf. note 3, above). On the contrary, no less a figure than Norbert Wiener, the father of cybernetics, centrally drew on virtue ethics in his *The Human Use of Human Beings: Cybernetics and Society* (1954). Wiener's volume is widely regarded a foundational in establishing what became computer ethics and its subsequent offshoots: hence, virtue ethics is all but built in from the beginning of these domains. Moreover, virtue ethics has enjoyed a rather remarkable renaissance over the past two decades or so, most especially within the areas of concern in play here, as instantiated by Vallor (2009, 2011a, b, 2015, cf. Spiekermann in press). For an overview of virtue ethics in Information and Computing Ethics and Media and Communication Studies, see Ess (2015).

developed new virtues uniquely important for its own flourishing and contentment (*eudaimonia*) as an artificial ethical being (Sullins, personal communication).

Love and Sex: Not Just for Humans Anymore?

To state the obvious, attention (especially male attention) to developing robots that might satisfy our erotic and sexual desires and interests has a long, long history. Sullins notes Ovid's story of Pygmalion, the sculptor who created a female statue so beautiful that he lost interest in human women, and whose greatest wish, to have the statue come alive, was granted by Aphrodite (2012, p. 398). The Pygmalion theme continues through various versions of fembots, sexbots, female AIs, etc. that have populated movies from *Metropolis* to *I, Robot* to *Ex Machina*: a female robot that can serve as one's lover, if not sex slave, appears to be of central interest.

And of course, love and sex with robots—real or yet to come—have been extensively explored. One important point of reference here is David Levy, whose *Love and Sex with Robots: The Evolution of Human–Robot Relations* (2007) sought to argue that human beings would indeed come to enjoy loving sexual relationships with social robots—at least, presuming that there were sufficient advanced as to rather closely imitate a human lover in important ways. More recently, John Sullins has criticized Levy's arguments (and their kin)—in good measure, as based on a virtue ethics approach (2012). Two of Sullins' points are especially relevant here. First, there appears to be considerable consensus that we cannot build AIs of sufficient sophistication as to be capable of a first-person experience of *emotion*. Given this, the approach is rather to develop robots that can *imitate* emotion—so-called *artificial emotion*. The game here is to simulate for a human being the embodied externalities or appearances affiliated with a given emotion: as is now well established, the simulacrum of emotion usually succeeds in triggering a emotive response from human beings in turn. For Levy, the appearance of love from a robot is sufficient because it thereby creates in the human (apparently) the same sensibility of being loved; for Sullins, this is an unethical trick. So he notes: "It is unethical to play on deep-seated human psychological weaknesses put there by evolutionary pressure as this is disrespectful of human agency" (Sullins 2012, p. 408).

Secondly, Sullins draws on Plato's discussion of *eros* in the *Symposium* to argue that human erotic love involves three key elements: (1) our <u>not</u> knowing fully what we need or desire in a lover, in part because (2) a human lover, as enjoying his or her own autonomy is thereby, as I would put it in Levinas' term, an Other, i.e., a person irreducibly *different* from us (1987).[6] This difference manifests itself as the Other brings into our relationship his or her own interests, desires, experiences, fallibilities, strengths, demands, etc. This means first of all

---

6 For a strong defense of the Otherness of social robots, as partly grounded in Levinas, (cf. Gunkel 2014, Sandry 2015).

that the Other is *not* a being I can construct and thus fully control. This further means—reinforcing our own not knowing fully what we desire or need in erotic love—that the erotic relationship essentially includes all the possibilities of surprise, resistance, unexpected gifts and disappointments, etc. that are necessary outcomes of engagement with a genuinely autonomous Other.

Another way of stating these points is that such an erotic relationship cannot be coerced. As we learn with especial painfulness in our first longings for and meanderings through erotic love in adolescence—such love cannot be evoked or forced into coming into being at will; nor can we fully control such a relationship at every moment. The latter might be possible with slaves, including robots as slaves: manifestly, this control and responsiveness to our wills are primary drivers in the fascination with (literally) man-made lovers, from Pygmalion to Ava in *Ex Machina*. Equally manifestly, however, such control and demand for compliant responsiveness is not possible with autonomous human beings who sustain their Otherness. Taken together, these first two points lead to the third dimension of *eros*: engaging in and sustaining relationship with the erotic Other will inevitably impose demands on us—including the demands and disappointments that serve our development as virtuous human beings.

Sullins invokes virtue ethics here, in part as he draws on the work of Mark Coeckelbergh, who asks how far robot design will lead to a good life, a life of flourishing (Coeckelbergh 2009, cited in Sullins 2012, p. 402). Most briefly, virtue ethics argues that we require certain virtues (habits, practices, or excellences) in order to pursue and attain a good life, a life of flourishing—where friendship is the primary example of a human relationship that requires and fosters such virtues (Ess 2013, pp. 238ff.). As Shannon Vallor (2009, 2011b) points out, such virtues include empathy, patience, and perseverance—abilities that do not come naturally or easily. On the contrary, acquiring these virtues is *difficult*—and so require practice, and so require relationships that thus push us to take up and practice what we might otherwise cheerfully avoid.

In these terms, then, erotic love with human Others are primary instantiations of relationships that are virtuous in the specific sense that they thereby push us beyond what we would be, so to speak, on our own, left to our own devices. Sullins makes this point as follows:

> We don't need just a machine to have sex with, we need one that makes us and the robot better by being with one another. We will have achieved nothing of moral worth by building machines that provide us with less as they will distract us from the more valuable pursuit of the kind of love that will expand our moral horizons through the experience of authentic love. (2012, p. 405)

That is, virtue ethics thus highlights a key failure of sex with robots. Such sex—however good and pleasant it might be on other grounds—does not fulfill a central feature of erotic love: engaging with the Other as Other—the Other whose erotic attraction for us thereby compels us to remain in relationship, to learn the virtues that will sustain and enhance relationship. Human *eros*, as drawing together two such Others, is thus a central way in which we thereby become better, expanding "our moral horizons" precisely as we thereby become more in and through such relationship than we would otherwise be (cf.Vallor 2011a). So Sullins concludes:

> [… T]he main lesson Socrates was trying to give us in the *Symposium* is that we come into a relationship impoverished, only half knowing what we need; we can only find the philosophically erotic through the encounter with the complexity of the beloved, complexity that not only includes passion, but may include a little pain and rejection from which we learn and grow. (2012, p. 408)

In sum: however sophisticated our social robots may become, so long as they are creatures we have explicitly and intentionally designed to meet what we think our needs may be; so long as they themselves cannot experience emotion, but can only trick us into thinking that they do; and so long as they are not full-fledged Others (but only capable of an artificial agency and autonomy)—our relationships with them, whatever other goods and benefits they may bring, thus cannot count as either fully erotic or fully virtuous in the senses developed here.

Good Sex—and Complete Sex

We can reinforce and expand Sullins' approach in a number of ways—beginning, for example, with Rollo May's analyses of *eros* and sexuality in the modern world. Writing at the height of the 1960s' sexual revolution, May captures much of the contrast underlying Sullins' analysis in his summary sentence, "We fly to the sensation of sex in order to avoid the passion of eros" (1969, p. 65).

Moreover, Sara Ruddick (1975) offers a phenomenological analysis of diverse sexual experiences that helps illuminate still more specific and critical conditions of human love and sexuality—ones constituting what she calls "complete sex" that further connect with both Kantian deontology and virtue ethics. A careful review of Ruddick's account will powerfully enhance Sullins' attention to erotic love as a critical site of difference between human and machine.

Ruddick draws from notions of completeness articulated by Sartre, Merleau-Ponty, and Thomas Nagel, where completeness depends first of all on *embodiment* (1975, p. 88). Completeness in sex occurs "when each partner's embodying desire is active and actively responsive to the other's" (1975, p. 89). Somewhat more fully:

> The partner *actively* desires another person's desire. Active desiring includes more than embodiment, which might be achieved in objectless masturbation. It is more, also, than merely being aroused by and then taken over by desire, though it may come about as a result of deliberate arousal. It commits the actively desiring

person to her desire and *requires her to identify with it*—that is, *to recognize herself as a sexual agent* as well as respondent. (ibid., emphasis added)

Moreover, this embodied sexual agency is mutual:

> In complete sex, desire is recognized by a responding and active desire that commits the other, as it committed the partner. Given responding desire, both people identify themselves as sexually desiring the other. (ibid.)

As she goes on to emphasize:

> [… I]n complete sex two persons embodied by sexual desire actively desire and respond to each other's active desire. … complete sex is reciprocal sex. The partners, whatever the circumstances of their coming together, are equal in activity and responsiveness of desire. (ibid., p. 90)

To say this somewhat differently: in complete sex, our desires are not simply for the Other—in addition, we desire that the other desire our desire. We desire to be fully desired as the full and complete embodied persons that we are—including our specific desires for the Other and his/her desire for us. (A Heideggerian might be pleased with the phrase: in complete sex, desire desires (the) desiring (of the Other).)

In short, complete sex as described here requires fully embodied agency, infused with mutual desire, in a relationship marked by equality and reciprocity. In particular, complete sex in these ways depends specifically upon our conscious awareness of both our own desire and that of the Other as mutual and mutually reinforcing.

First of all, Ruddick's account thus works to counter then-prevailing Cartesian dualisms—i.e., views that presumed a strong ontological split between mind (as *res cogitans* for Descartes) and body (as *res extensa*). Such dualisms, we should note, were strongly in play in the emergence of early approaches to AI and, as we have seen, were later countered with AI and cognitive science precisely by phenomenological insights drawn from Merleau-Ponty and others (see above, p. 75). At this point we can notice: if we do assume such a strong mind-body split, then, as Descartes made clear centuries ago, all that we count as important and valuable about the human *person* can only reside in the mind: the body, as part of the larger natural order, can only be matter—where matter and nature for Descartes are thereby made the legitimate objects of our "mastery and possession" (Descartes 1972, p. 119). Sex and sexuality, it follows, can only be matters of physical stimulus and response—certainly pleasant and enjoyable enough under salutary circumstances: but it is difficult to see how a Cartesian ontology of the human person could thereby connect sexuality and body, on the one hand, with the cluster of values and norms that many of us associate with the human person, on the other—beginning with our sense of individual or unique personhood, and our sense that we are to be respected as persons. On the contrary, it would seem that the default consequence of Cartesian dualism for sexuality is that bodies are "just meat"—i.e., objects and agents of sex as solely physical. (Indeed, as many Second Wave feminists have carefully documented and demonstrated, early modern understandings of reason and rationality all too often fell into an easy dichotomy of males as rational and thus free persons, vs. females as literally embodying nature, feeling, and sexuality—making modernist efforts to understand women as somehow *persons* worthy of respect and emancipation philosophically contorted if not simply impossible (e.g. Porter 2003, pp. 257ff.).

Ruddick takes up embodiment as countering such dualisms. To be sure, we have experiences of distance from our own body, of observing or somehow using them to fulfill specific intentions. Nonetheless, Ruddick points out that "On some occasions, however, such as in physical combat, sport, physical suffering, or danger, we 'become' our bodies; our consciousness becomes bodily experience of bodily activity" (1975, p. 88). To be sure, Ruddick is not alone on this point. Prior to Ruddick, for example, the phenomenologist Maurice Natanson observed:

> I am neither "in" my body nor "attached to" it; it does not belong to me or go along with me. *I am my body.* There is no distance between my hand and its grasping. [...] Instead of the common-sense way of thinking of the body in space at some time, I am a corporeality Here and Now whose being in the world is disclosed to me as mine. (1970, p. 11)

As an extension of this phenomenological anti-dualism, the German phenomenological philosopher Barbara Becker (2001) developed the neologism *LeibSubjekt*—"body-subject"—in an effort to provide us with an alternative term for the self-body as an inextricable unity that, again, would help us both conceptually and linguistically overcome deeply entrenched Cartesian dualisms and affiliated terminologies.

Given this account of complete sex, Ruddick argues that complete sex is beneficial in instrumental ways—first of all, as "they are conducive to our psychological well being ... " (1975, p. 97). Moreover, the mutual responsiveness of complete sex "satisfies a general desire to be recognized as a particular person and to make a difference to other particular 'real' people" (ibid.).

Even better, Ruddick can now show that complete sex acts enjoy a moral preference—and this for three reasons:

> They tend to resolve tensions fundamental to moral life; they are conducive to emotions that, if they become stable and dominant, are in turn conducive to the virtue of loving; and they involve a preeminent moral virtue—respect for persons. (ibid., p. 98)[7]

---

7   It is important to note that Ruddick does not intend to argue that all of our sexual experiences must meet these conditions and thus fulfill these moral norms (1975, p. 101).

Elaborating on her first point, Ruddick points out "morality" is often what is civilizing, social, and regulating: desire—especially sexual desire—is discontent that resists regulation. In complete sex, however, "Mutually responding partners confirm each others' desires and declare them good," thus helping to overcome the otherwise usual tension between our social and private lives (ibid., p. 98).

Secondly, Ruddick observes that complete sex acts are likely to lead to affection, as well as feelings such as "gratitude, tenderness, pride, appreciation, dependency, and others" (ibid., pp. 98f.). Specifically:

> These feelings magnify their object who occasioned them, making him *unique* among men. When these magnifying feelings become stable and habitual they are conducive to love ... of a particular sexual partner. (ibid., p. 99, emphasis added)

Against the potential objection that such particularly focused love may seem selfish and thus limited in comparison with a more universal love, Ruddick counters: "However, even 'selfish' love is a *virtue*, a disposition to care for someone as her interests and demands would dictate" (ibid., emphasis added).

I take it that loving counts as a virtue here in part because it is understood as not simply an emotion or a passion (something that happens to us, out of our control)—but as a practice, one fostered, e.g., by ethical commitments of fidelity to the Other. In particular, we are not always naturally or easily inspired to remain in relationship with an Other at all times. Specifically, it is perhaps the commonest of relational experiences to recognize that sexual passion and attraction for a specific person will ebb and flow; and it is most tempting to break off especially after periods of stressed and strained relationships, e.g., as often brought on by the burdens of parenthood, financial difficulties, etc. But loving as a practice—as a capacity or habit of regarding the Other as one's partner nonetheless—along with other virtues critical to friendship such as empathy, patience, and perseverance (Vallor 2009, 2011b), may also serve to help us through such difficult times.[8] Indeed, passion and attraction may return thereafter in new, deeper, and richer ways, such that relationship is easier and more rewarding. But to have such experience of relationship requires loving as a practice, as a virtue—an engagement of commitment and work, exactly when we are least inclined by nature or emotion to carry on. Lastly, Ruddick follows Sartre's suggestion that complete sex acts entail:

Rather, she seeks to highlight our sexual experiences, when they do engage us as such body-subjects, as thereby immediately bringing in their train these important ethical components.

8 In Danish one helpfully speaks idiomatically of going "fra forelskelse til kærlighed"—roughly, from falling in love to loving. "Forelskelse" refers more to the initial, giddy, passionate experiences surrounding falling in love and new love, where kærlighed has connotations of long-term relationship and deep commitment—what endures nonetheless in the face of changing passion and difficulties.

... a respect for persons. Each person remains conscious and responsible, a 'subject' rather than a depersonalized, will-less, or manipulated 'object.' Each actively desires that the other likewise remain a 'subject'. (1975, p. 99)

We can recognize here an originally Kantian insistence on respect for *persons* as autonomous and equal: such an insistence is clearly consistent with the requirements of mutuality and respect Ruddick limns out—and in fact, Ruddick explicitly invokes autonomy in a later discussion, as we will soon see. At the same time, however, Ruddick identifies such respect as "a central *virtue* when matters of justice and obligation are at issue" (ibid., p. 99, emphasis added). This respect, moreover, "requires that *actual present* partners participate, partners whose desires are recognized and endorsed" (ibid., p. 100). To say this differently, we must be aware of the desiring presence—what might otherwise be called shared attention (Broadbent and Lobet-Maris 2014)—of the Other, an attention that carries with it respect for one another as equals. As Ruddick points out, such mutual respect is especially at risk precisely in the context of sexuality:

> Respect for persons typically requires taking a distance from both one's own demands and those of others. But in sex acts the demands of desire take over, and equal distance is replaced by mutual responsiveness. Respect typically requires refusing to treat another person merely as a means to fulfilling demands. In sex acts, [however] another person is so clearly a means to satisfaction that she is always on the verge of becoming merely a means ('intercourse counterfeits masturbation'). In complete sex acts, instrumentality vanishes only because it is mutual and mutually desired. Respect requires encouraging, or at least protecting, the *autonomy* of another. In complete sex, autonomy of will is recruited by desire, and freedom from others is replaced by frank dependence on another person's desire. Again the respect consists in the reciprocity of desiring dependence, which bypasses rather than violates autonomy. (Ruddick 1975, p. 100, emphasis added)

In sum, insofar as complete sex engages us not simply as bodies but as whole persons—the ineluctable presence of the *person* engages our sense of the Other as not solely the object of sexual desire, but as an Other as a distinct and unique person first of all, with whom I stand in a distinct and unique relationship of mutuality and reciprocal care and concern. Complex sex is sexuality inextricably interwoven and suffused with our distinctive identities as persons: such sexuality thereby embodies the felt uniqueness of our relationship, infused with other feelings such as tenderness, pride, gratitude, appreciation, etc. (ibid., p. 98). All of this reinforces our experience of the Other as a person, not as a thing—and thereby invokes first of all a Kantian duty of respect for the Other as a person, i.e., as an autonomous and unique entity. Complete sex thus stands as a counterpart to more dualistic approaches that make it difficult not to treat a sex partner as anything but an object, i.e., as a thing divorced from personhood. Ruddick's analysis thereby articulates what I think many of us find most valuable—indeed, definitive of ourselves and our identity as complete human beings—namely, the sense of being loved fully and completely, precisely as the unique BodySubjects that we experience ourselves to be much of the time. Perhaps most centrally, this experience of affirming and being affirmed as whole persons specifically requires the mutuality of desire—and in this way, complete sex is rather remarkable—and for a strict Cartesian, conceptually impossible: *contra* the strong tendencies of dualistic accounts to denigrate, if not demonize body and sexuality (and, correlatively, women, as the orthodox tradition of Original Sin makes most explicit: Ess 2014), complete sex not only highlights the central importance of sexual desire to the affirmation of our personhood as whole, embodied beings—it further establishes complete sex as an anchor for the critical virtues of respect, equality, and loving.

What's Complete Sex Got to Do with It? Love and Friendship between Humans and Machines

Ruddick's account of complete sex thus powerfully undergirds and complements Sullins' account of erotic love as a relationality that we enter into with only an incomplete understanding of our own desires and interests—a relationality that, as fully preserving the Other as an autonomous and independent entity beyond our design and control, thus allows the Other to call us into an unfolding of our best potentials as these contribute to good lives of flourishing. Ruddick makes more explicit how complete sex—or, we may be justified in saying, complete *eros*—can do so: as engaging us as BodySubjects, as Others in relationships of mutuality of desire and care, complete sex/*eros* thus sustains the presence of the Person as an autonomy deserving respect as an equal. At the same time, complete sex/*eros* entails loving as a virtue: as we practice and become better at loving the Other as Other—in conjunction with other virtues such as patience and perseverance—we thereby unfold these specific virtues as forms of flourishing and components of good lives.

In this light, Sullins' critiques of what we can think of as strong continuity approaches to love and sex with robots (such as Levy's) are likewise enhanced and complemented.

Again, complete sex/*eros* first of all requires the presence of a person as an Other—an autonomy in the full-fledged human sense, in contrast with what Sullins later calls artificial autonomy. More broadly, complete sex/*eros* thus requires the engagement of the Other as a BodySubject—where the subject enjoys the full capacities of human awareness, emotion, and desire. What might sex and sexuality with a social robot look like *vis-à-vis* these criteria? Let us optimistically assume with Sullins (and others, e.g. Weckert 2011) that we will be soon be able to construct a social robot that enjoys such critical characteristics as: artificial autonomy, artificial *phronesis*, artificial agency and responsibility. It is by no means clear, as we have seen, that such a robot will enjoy awareness or self-consciousness as we construe it (Searle 2014)[9]; it seems rather clear, at least this point, that such a robot will only be capable of evincing artificial emotions—i.e., enacting behaviors that can trick us into believing the robot cares for us.

To begin with, nothing in these accounts forbids or denigrates the use of sexbots as beneficent and morally legitimate partners for "just sex," i.e. sexual experiences such as masturbation, those contracted with a prostitute, and/or those undertaken within a clear understanding of their limits, e.g., one-night stands. But we can now see that however enjoyable sex might be with such a social robot, it will not fulfill the criteria of complete sex/*eros*. First of all, artificial autonomy is not human autonomy—and so it is difficult to see how such a social robot would qualify as an Other in the strong sense required by complete sex/*eros*. Secondly, it is not solely the case, as Sullins puts it, that artificial emotions are an unethical trick (2012, p. 408). In addition, the absence of a sense of selfhood, coupled with a lack of awareness of emotions—specifically sexual desire—short-circuits any possibility of the experience of mutuality of desire. However skills and pleasing my robot lover might be—as an entity incapable of awareness and desire, it cannot be aware of desiring me, much less of desiring my desire. In turn, it can never give me the vital experience of knowing that it is an Other who desires my desire. Absent these, it cannot provide me with the fullest affirmation of my sense of self as a unique person, as this selfhood includes my desire that my desire be desired. And without these key dimensions of complete sex or *eros*, the robot lover will also fail to anchor and inspire the development of the primary virtues affiliated with complete sex—respect for persons, equality, and loving.

That is, a virtue ethics approach would have us ask: what virtues are fostered and/or hindered through sex with such a social robot? As Sullins has argued, it may well be possible and desirable to build such a social robot as a creature capable of virtues—and so up to a point, at least, such a robot could engage with us in relationships that might foster important virtues on both sides. Indeed, it is conceivable that as social robots develop as virtuous agents, they might well develop distinctive virtues that would in turn challenge us to develop new virtues that would further enhance our capacities for flourishing and good lives. But however important and beneficent such relationships might be—they would seem to again fall short *vis-à-vis* how the virtues of mutual respect, equality, and loving

---

9   Indeed, Selmer Bringsjord is emphatic on this point: he has repeatedly argued that " ... genuine phenomenal consciousness is impossible for a mere machine to have, and true self-consciousness would require phenomenal consciousness" (Bringsjord et al. 2015, p. 2). He goes on to point out that "Nonetheless, the logico-mathematical structure and form of self-consciousness can be ascertained and specified, and these specifications can then be processed computationally in such a way as to meet clear tests of mental ability and skill. In short, computing machines, AIs, robots, etc. are all "zombies," but these zombies can be engineered to pass tests" (ibid.). In these terms, you could have good sex with a robot zombie, perhaps—but not complete sex as the latter requires self-consciousness as the foundation of autonomy, respect for personhood, and mutuality of desire.

are fostered within relationships of complete sex/*eros*. First of all, without full autonomy, personal awareness, and emotions (including those affiliated with embodiment and sexuality)—it is not clear how they could engage us in the same way as a human Other. Especially as we would be aware that their autonomy is artificial autonomy, I do not see that such robots would enjoin us to pursue and sustain relationships of respect and equality. By the same token, as we would be aware that their "emotions" are artificial emotions and thus appearance (if not simple trickery), I do not see how sexual engagement with such robots would issue in the same experiences of complete sex/*eros*—most especially the sense of being loved and *desired* wholly and completely for who I am as an embodied person.

Indeed, the absence of these two conditions would be amplified by a third awareness: the social robot is a creature of human design and construction, one that can be bought and sold at market. In sharp contrast with how loving relationships with human beings can, as undergirded by ethical commitments to one another as human beings and as conjoined with such virtues as patience and perseverance, inspire me to stay within such a relationship despite its rough edges and patches, and thus unfold my full humanity even further—it would seem that if a relationship with a social robot somehow became boring, unpleasant, much less nasty, the easiest and most straightforward thing would be return it to the manufacturer or sell it as a used artifact. This possible response to a sexbot no longer of interest to us does not appear to nurture the virtues of empathy, mutuality, etc.: rather, it would seem to make these virtues superfluous (cf. Vallor 2011a, 2015).

Concluding Remarks

In sum, however far we may come with developing social robots capable of instantiating and replicating important human capacities—including artificial autonomy, artificial *phronesis*, and artificial emotions—sexual engagements with such robots, however good and beneficent they might be on other grounds, will not constitute the sorts of human relationships we have explored in terms of complete sex/*eros*.

First of all, I hope that these insights and arguments will thus contribute to the ongoing discussions and debates within social robotics and roboethics, and, more generally, to the ongoing developments within cognitive science, especially as informed by phenomenology. Most broadly, as these explorations help further sharpen and elaborate our understanding and sensibilities as to what machines can do, and what, so far at least, remains distinctively human, they thus stand within philosophical explorations of the computational turn. To state the point most sharply: what remains human first includes our capacities for strong autonomy, *phronesis*, embodied emotions—including the emotions clustering about sexual desire and complete sex. And, while social robots could conceivably contribute in important ways to our pursuit of good lives and flourishing by way of developing the requisite virtues thereto—I have argued that social robots, as lacking strong autonomy, genuine emotion, and (so far) self-awareness as an embodied being with such capacities, thereby fall short of what is required in complete sex to establish experiences of mutual desire and respect, and thereby social robots will not likely foster the virtue of loving as a key virtue for flourishing and a good life.

More particularly, the insights developed here through careful attending to what counts as complete sex/*eros* would appear to support weak continuity approaches. But the conjunction with virtue ethics, however salutary for these purposes, may not be good news for those who wish to insist still more strongly on the uniqueness or distinctive of human beings vs. robots. Rather, setting up Just War theory and *phronesis* as standards for ethical judgment as a condition of ethical responsibility is to set a very high bar indeed. Specifically, Just War theory justifies the right of conscientious objection—of principled disobedience that rests in the capacity to judge when a superior order (e.g., to kill civilians indiscriminately) is *not* a legal order (because it violates the immunity of non-combatants). More broadly, civil rights activists such as Gandhi and Martin Luther King, Jr., called for civil disobedience to what they characterized as unjust laws—where the appeal to justice requires judging that certain criteria (e.g., equality) outweigh extant legislation (King 1964). To be sure, there are spectacular examples of such disobedience in history—as in the resistance of a few Americans to the orders and actions of their comrades at My Lai (cf. Vallor 2015, p. 115) and in the civil rights movements in India, the United States, and elsewhere. But it would seem that most of us prefer to obey orders most of the time.

By the same token, the criteria of complete sex/*eros* as normative criteria for sexual relationships likewise set up a very high bar. Again, the virtues that come into play here—mutual respect, equality, loving—are neither easy nor natural: as Ruddick indicates, and as Sullins and Vallor reinforce regarding affiliated virtues, these are difficult, especially at the beginning, and it is tempting to put practice aside and be satisfied with what comes more easily. In these directions, Sherry Turkle has most famously argued that in fact we turn to the ease and convenience of electronic communication and robots in order to avoid more directly human relationships—with the result that we are "alone together," i.e., increasingly incapable of developing and sustaining fully human relationships with all their difficulties and intractable demands (2012, cf. Vallor 2011a, Vallor 2015).

In this light, a last consequence of taking up a virtue ethics approach here is to thereby recognize the often uncomfortable demands such an ethics places upon us. Most briefly, we are not simply human beings by virtue of being born with human DNA. Rather, we become human beings—or not—in some measure as we actively pursue and practice the virtues arguably requisite for good lives of flourishing. To put the point sharply: if we are not to be beings fully replicable and thus replaceable by social robots—it would appear that we are thus required to take up the difficult and often uncomfortable chores of acquiring and practicing such virtues as patience, perseverance, empathy, mutual respect, equality, and loving. This means most foundationally pursuing the practice of *phronesis*—including its capacity to lead us not only to relationships of complete sex/*eros*, but also, perhaps inevitably, to moments of principled disobedience. As we have seen, both are difficult: we can add here, the latter can be fatal.

Failing to set these high bars, however, would seem to make it easier to conflate ourselves with sophisticated social robots. Some, perhaps many, especially within social science and natural science frameworks, may well believe that we are indeed such creatures: indeed, given the materialistic and deterministic foundations that define these frameworks, human beings could hardly be understood otherwise. From these and other perspectives (e.g., postmodernisms of various sorts), all discussion of autonomy, mutuality, etc. are but the quaint artifacts of less enlightened ages. I rather think, however, that human beings remain capable of choice; and I hope this essay might inspire some of us to choose more to pursue the virtues discussed here, and less to put them aside for the sake of what is easier and more convenient. This is a high bar—but a possible one. I would also hope that the prospect of good lives, of flourishing—specifically, by seeking to become better human beings by acquiring the virtues that will help us become better friends and lovers—might be sufficient inspiration for undertaking these arts of being human. In this way, our efforts to build better robots might evoke renewed resolve to become better human beings as well.

References

Arendt, H. 1972. *Crises of the Republic*. New York: Harcourt, Brace, Javonovich.

Becker, B. 2001. "The Disappearance of Materiality?" In *The Multiple and the Mutable Subject*, edited by V. Lemecha and R. Stone, 58-77. Winnipeg: St. Norbert Arts Centre.

Boden, M. 2006. *Mind as Machine: A History of Cognitive Science*. Oxford: Clarendon Press.

Bringsjord, S., J. Licato, N.S. Govindarajulu, R. Ghosh, and A. Sen. 2015. "Real Robots That Pass Human Tests of Self-Consciousness." Proceedings of RO-MAN 2015 (The 24th International Symposium on Robot and Human Interactive Communication), Kobe, Japan, August 31-September 4, 2015.

Broadbent, S. and C. Lobet-Maris. 2014. "Towards a Grey Ecology." In *The Onlife Manifesto: Being Human in a Hyperconnected Era*, edited by Luciano Floridi, 111-24. Springer International Publishing. doi: http://dx.doi.org/10.1007/978-3-319-04093-6_15.

Coeckelbergh, M. 2009. "Personal Robots, Appearance, and Human Good: A Methodological Reflection on Roboethics." *International Journal of Social Robotics* 1(3): 217-21. doi: http://dx.doi.org/10.1007/s12369-009-0026-2.

Coeckelbergh, M. 2012. *Growing Moral Relations: Critique of Moral Status Ascription*. London: Palgrave Macmillan.

Descartes, R. 1972. "Discourse on Method." In *The Philosophical Works of Descartes*, translated by E.S. Haldane and G.R.T. Ross, 81-130. Cambridge: Cambridge University Press. Original edition, 1637.

Dreyfus, H.L. 1972. *What Computers Can't Do: A Critique of Artificial Reason.* New York: Harper & Row.

Dreyfus, H.L. 1992. *What Computers Still Can't Do: A Critique of Artificial Reason.* New York: MIT Press.

Ess, C.M. 1983. *Analogy in the Critical Works: Kant's Transcendental Philosophy as Analectical Thought.* Ann Arbor, MI: University Microfilms International.

Ess, C.M. 2009. *Digital Media Ethics.* Oxford: Polity Press.

Ess, C.M. 2013. *Digital Media Ethics.* 2 ed. Oxford: Polity Press.

Ess, C.M. 2014. "Ethics at the Boundaries of the Virtual." In *The Oxford Handbook of Virtuality*, edited by Mark Grimshaw, 683-97. Oxford: Oxford University Press. doi: http://dx.doi.org/10.1093/oxfordhb/9780199826162.013.009.

Ess, C.M. 2015. "The Good Life: Selfhood and Virtue Ethics in the Digital Age." In *Communication and the "Good Life"*, edited by Helen Wang, (ICA Themebook, 2014), 2017-29. New York: Peter Lang.

Gerdes, A. 2014. "Ethical Issues Concerning Lethal Autonomous Robots in Warfare." In *Sociable Robots and the Future of Social Relations: Proceedings of Robo-Philosophy 2014*, edited by Johanna Seibt, Raul Hakli and Marco Nørskov, 277-89. Amsterdam: IOS Press Ebooks. doi: http://dx.doi.org/10.3233/978-1-61499-480-0-277.

Godfrey-Smith, P. 1994. "Spencer and Dewey on Life and Mind." In *Artificial Life 4*, edited by R. Brooks and P. Maes, 80-89. Cambridge MA: MIT Press.

Gunkel, D. 2014. "The Other Question: The Issue of Robot Rights." In *Sociable Robots and the Future of Social Relations: Proceedings of Robo-Philosophy 2014*, edited by Johanna Seibt, Raul Hakli and Marco Nørskov, 13-14. Amsterdam: IOS Press Ebooks. doi: http://dx.doi.org/10.3233/978-1-61499-480-0-13.

Kavathatzopoulos, I. 2014. "Independent Agents and Ethics." *ICT and Society: IFIP Advances in Information and Communication Technology* 431: 39-46.

Kavathatzopoulos, I. and R. Asai. 2013. "Can Machines Make Ethical Decisions?" *Artificial Intelligence Applications and Innovations: IFIP Advances in Information and Communication Technology* 412: 693-9.

King, M.L., Jr. 1964. "Letter from the Birmingham Jail." In *Why We Can't Wait*, edited by Martin Luther King, Jr., 77-100. New York: Mentor. Original edition, 1964.

Levinas, E. 1987. *Time and the Other and Additional Essays.* Translated by Richard A. Cohen. Pittsburgh, PA: Duquesne University Press.

Levy, D. 2007. *Love and Sex with Robots: The Evolution of Human–Robot Relationships.* New York: Harper Collins.

May, R. 1969. *Love and Will.* New York: W.W. Norton and Co.

McDermott, D. 2008. "Why Ethics Is a High Hurdle for Ai." North American Conference on Computers and Philosophy (NA-CAP), Bloomington, Indiana, July, 2008.

Natanson, M.A. 1970. *The Journeying Self: A Study in Philosophy and Social Role.* Reading, MA: Addison-Wesley.

Porter, R. 2003. *Flesh in the Age of Reason: The Modern Foundations of Body and Soul*. New York: W.W. Norton & Co.

Ruddick, S. 1975. "Better Sex." In *Philosophy and Sex*, edited by Robert Baker and Frederick Elliston, 280-99. Amherst, NY: Prometheus Books.

Sandry, E. 2015. "Re-Evaluating the Form and Communication of Social Robots: The Benefits of Collaborating with Machinelike Robots." *International Journal of Social Robotics* 7(3):335-46. doi: http://dx.doi.org/10.1007/s12369-014-0278-3.

Searle, J.R. 2014. "What Your Computer Can't Know." The New York Review of Books. Accessed May 22, 2015. http://www.nybooks.com/articles/archives/2014/oct/09/what-your-computer-cant-know.

Seibt, J. 2014. "Introduction." In *Sociable Robots and the Future of Social Relations: Proceedings of Robo-Philosophy 2014*, edited by Johanna Seibt, Raul Hakli and Marco Nørskov, vii-viii. Amsterdam: IOS Press Ebooks.

Spiekermann, S. In press. *Ethical It Innovation: A Value-Based System Design Approach*. New York: Taylor & Francis.

Sullins, J.P. 2012. "Robots, Love, and Sex: The Ethics of Building a Love Machine." *IEEE Transactions on Affective Computing* 3(4): 398-409. doi: http://dx.doi.org/10.1109/t-affc.2012.31.

Sullins, J.P. 2014. "Machine Morality Operationalized." In *Sociable Robots and the Future of Social Relations: Proceedings of Robo-Philosophy 2014*, edited by Johanna Seibt, Raul Hakli and Marco Nørskov, 17-17. Amsterdam: IOS Press Ebooks. doi: http://dx.doi.org/10.3233/978-1-61499-480-0-17.

Turkle, S. 2012. *Alone Together: Why We Expect More from Technology and Less from Each Other*. New York: Basic Books. Original edition, 2011.

Vallor, S. 2009. "Social Networking Technology and the Virtues." *Ethics and Information Technology* 12(2): 157-70. doi: http://dx.doi.org/10.1007/s10676-009-9202-1.

Vallor, S. 2011a. "Carebots and Caregivers: Sustaining the Ethical Ideal of Care in the Twenty-First Century." *Philosophy & Technology* 24(3): 251-68. doi: http://dx.doi.org/10.1007/s13347-011-0015-x.

Vallor, S. 2011b. "Flourishing on Facebook: Virtue Friendship & New Social Media." *Ethics and Information Technology* 14(3): 185-99. doi: http://dx.doi.org/10.1007/s10676-010-9262-2.

Vallor, S. 2015. "Moral Deskilling and Upskilling in a New Machine Age: Reflections on the Ambiguous Future of Character." *Philosophy & Technology* 28(1): 107-24. doi: http://dx.doi.org/10.1007/s13347-014-0156-9.

Wallach, W. 2014. Moral Machines and Human Ethics. Presentation, Robo-Philosophy 2014: Socialbe Robots and the Future of Social Relations. Aarhus University, Denmark.

Wallach, W. and C. Allen. 2009. *Moral Machines: Teaching Robots Right from Wrong*. New York: Oxford University Press.

Weckert, J. 2011. "Trusting Software Agents." In *Trust and Virtual Worlds: Contemporary Perspectives*, edited by C. Ess and M. Thorseth, 89-102. New York: Peter Lang.

Weizenbaum, J. 1976. *Computer Power and Human Reason: From Judgment to Calculation*. New York: W.H. Freeman.

Wiener, N. 1954. *The Human Use of Human Beings: Cybernetics and Society*. Garden City, NY: Doubleday Anchor. Original edition, 1950.

Wilson, R.A. and L. Foglia. 2011. "Embodied Cognitions." In *The Stanford Encyclopedia of Philosophy*, ed Edward N. Zalta. Fall 2011 Edition. http://plato.stanford.edu/archives/fall2011/entries/embodied-cognition.

PART II
Potential

Chapter 5

Ethics Boards for Research in Robotics and Artificial Intelligence: Is it Too Soon to Act?

John P. Sullins

As a condition of the sale of the AI company DeepMind to Google, DeepMind's founder Demis Hassabis stipulated that Google must set up an internal ethics board to review the research which the company does in the field of AI, and presumably robotics as well (Burrell, 2014). Negative reaction to this idea was quick with one unnamed machine learning researcher stating in an article on *Re/code* that such a move is unnecessary as "We're a hell of a long way from needing to worry about the ethics of AI" (Gannes and Temple 2014). Is it too early to start building a professional ethics regarding AI and robotics research? Would such a move only stifle research and fan media fears? Artificial Intelligence and Robotics have an image problem. When the machines and programs built by researchers in these fields fail, the media is quick to report that machines will never equal humans when it comes to thinking and acting intelligently in the real world. But when they show some success, the media is ever ready to pronounce the coming extinction of the human race at the hands of cold malevolent machine overlords. With hyperbole like this, it is no wonder that researchers are loath to put their experiments under scrutiny that might go public. Even given these problems, I will argue that the time has come for AI and robotics researchers to welcome the contributions to good product design that ethical review adds to the research and development process.

Introduction

There is nothing novel about forming internal ethics boards. There has been a practice in creating these entities that arguably got its start in medical review boards which began appearing in the early 1980s due to a desire to stave off malpractice claims. Many other industries also have ethics boards, and starting around the end of the 1990s, even at the highest levels in the corporate world, Chief Ethics Officers have begun to populate some executive boards (Clark 2006). Philosophers have an even longer tradition of arguing the valuable role that ethics can play in the design of technologies worth having for at least as far back as the turn of the last century, if not before. So why was it such big news when Google announced that it would form an internal ethics board as part of a deal to buy the AI startup company Deep Mind (Burrell 2014, Cohen 2014, Gannes and Temple 2014, Garling 2014, Prigg 2014, Weaver 2014)? The real story should be why Google did not already have an internal ethics review board and the embarrassing fact that it took the special pleading of Deep Mind's founder before they thought to do so, even though their products all have significant ethical impacts on their users. There does seem to be a board that has been formed now, but details on what it does and who is on it are sketchy (Lin and Selinger 2014). One of Google's vocal critics claims that this omission is due to a long standing aversion found in Google management to setting up internal rules or regulations that might restrict the creativity of its employees (Cleland 2014). The computer scientist Norman Matloff recently wrote that Silicon Valley executives have a particularly glaring problem when it comes to running ethical companies. He reports that ethical problems such as illegal wage fixing, disregard for the rampant privacy abuse caused by their products, and age discrimination are exacerbated by the "boy king" mentality of their founders, a kind of ethical immaturity (Matloff 2014). This may be true in general, but an additional concern that may be a factor in this is summed up by this anonymous quote: "We're a hell of a long way from needing to worry about the ethics of AI" (Gannes and Temple 2014). It makes some sense, why bother setting up new bureaucracy when there is nothing yet to monitor? In medicine we need people to monitor things like the ethics of end of life decision made by doctors because it is an everyday occurrence whereas AI only ever hurts people in movies and in fiction, it is just not a real concern.

Is it too early to start building a professional ethics regarding AI and robotics research? Would such a move only stifle research and fan media fears? To answer these questions we must acknowledge that artificial intelligence and robotics have somewhat of an image problem. When the machines and programs built by researchers in these fields fail, the media is quick to report that machines will never equal humans when it comes to thinking and acting intelligently in the real world. But when they show some success, the media is ever ready to pronounce the coming extinction of the human race at the hands of cold malevolent machine overlords. With hyperbole like this, it is no wonder that researchers are loath to put their experiments under scrutiny that might go public. But I will argue that the time has come for AI and robotics researchers to welcome the contributions to good product design that ethical review adds to the research and development process.

There are two key distinctions to be made before we get too much further with this argument. The argument against AI and robotics ethics boards largely turns on a failure to appreciate the distinction between AI entities and intelligent agents as well as the difference between human moral agents vs. artificial moral agents (AMAs). It is assumed there are no real ethical concerns raised by AI until we see full AI entities emerge, since this is decades, if not centuries away, so too is any need to worry about the ethics of AI.

At this time AI entities that match or supersede human intelligence are an idea, not a reality and they are only to be found in the works of science fiction. Intelligent agents are software or robotic systems that are capable of intelligently processing information or accomplishing tasks intelligently. There are a great many of these agents now in operation and more are added daily. Most of these are hardly noticeable as AI since they do not have personalities, desires nor emotions, no ability to do anything other than what they were programed to do. This distinction is often described in the literature of the philosophy of mind as hard AI vs soft AI.

Hard AI tries to emulate the human mind, whereas soft AI is just interested in engineering solutions to specific problems that require intelligence. We can design programs like Deep Blue that can beat chess grandmasters at their own game, but we have to remember that this very capable chess playing program does not even know what chess is or even that it was playing a game with another person. It was just an algorithm that, when placed in the right context, could win at chess. Deep Blue is an example of soft AI. Hard AI would not only know how to play chess but would have some other understanding of the world around it, for instance understanding what it meant to win or lose against a grand master.

There is another distinction to be made with artificial moral agents (AMAs), and human moral agents. As I have argued in other essays, AMAs do not have to be full AI entities to count as moral agents (Sullins 2006). Although I agree that it might be preferable for an AMA to be a full AI entity, but as of now nothing like that exists. There are some intelligent artificial moral agents that are designed to make decisions that have ethical impact on other humans. For instance, there are some that aid in making life and death decisions on the modern battlefield. Every one of these existing systems is implemented through intelligent agents, software that has no real idea what it is doing in a conscious or self-aware sense. As we will see in the argument that follows, it is a mistake to think that the ethical concerns raised by existing AI software are too new or experimental to require ethical analysis of their use and design.

AI and Robotic Ethics

There is a present need for ethics advising on AI and robotics research at the corporate level. Over most of human history, ethics has been seen as largely a religious or personal concern. It has only been in the last half century that professional business ethics has begun to take shape in reaction to the rough and tumble world of commerce. The free press has also grown in its ability to report on major abuses that used to go unnoticed in earlier centuries, and this coupled with new political systems that can call powerful individuals and organizations to account for their actions, has led to the rise of applied professional ethics. Political organizations now all have ethics committees, and committees and codes of ethics are commonly found in the majority of professional organizations and academic societies. But ethics as a professional concern in the modern research and development world has been a bit tardier.

There are some exceptions of course. Medicine has had the Hippocratic Oath for millennia and over the last century ethical codes have made even more significant inroads into curbing abuse in medical research as well as moderating research in the social sciences that involve human and/or animal subjects. Researchers in those fields are quite used to obtaining institutional review board (IRB) approval before they engage in studies that directly impact sentient subjects. Unless they have worked in interdisciplinary projects that required IRB approval computer scientists, programmers, and engineers are far less used to this process. This reluctance to seek IRB approval may have made sense to researchers in the early days of computer science who believed their investigations involved only closed systems that had very little effect outside of their own lab. That belief was misguided then, but is completely unsustainable now. Computer science is a victim of its own wild success, which means that their research is everybody's business now since their work has already fundamentally altered the lives of everyone on the planet. AI and robotics will be a change that will be at least as significant as the rise of the networked society, if not more so, which means that these innovations stand to have tremendous ethical impacts even in the short term (Sullins 2004, 2014).

The time has come for companies that engage in AI and Robotics research to do what is right and join the other professionals that take ethical review seriously.

Short Term Ethical Concerns

AI and robotics applications have a number of applications that are already raising ethical concerns in areas as diverse as the home, the workplace, medical, military, and police work. Even though we are not talking about robust AI entities, these small unconscious intelligent agents interact with our lives and that means they will need ethical and political policies in place to help develop these expanding relations in a positive way. In earlier work I have suggested that there are a number of open ethical issues when it comes to robotics that presently needs solutions (Sullins 2011). These ethical issues occur because all AI and robotics applications will record, gain access to, and synthesize information in ways that impact the lives of their owners and users. When I speak of accessing and synthesizing information I mean the automation of acting intelligently on information, for example as is done now in the automated making of credit decisions, design, and planning of new products, or even targeting decisions in warfare.

When it comes to the home, these applications will encounter the tricky ethical problems associated with privacy that we have seen with other information technologies, but this will be heightened as these applications will need to collect, store, and grant access to truly astonishing amounts of data regarding the daily lives of their users. In addition to this general concern we have to add the important subfield of roboethics that deals with the ethical problems that arise with robot or AI applications that are designed to aid in the care of animals, the elderly, or children in a home setting. AI or robotic devices that use affective computing[1] in order to interact with human users are a particularly troubling area. In order to do a psychological study in an academic institution utilizing children as subjects a research group would have to undergo rigorous IRB approval, but there seems to be nothing stopping a company that wanted to release an AI/robotic toy that manipulated children's' psychological tendencies through affective computing techniques designed to cause the children to love their toy. As long as the toy met the minimum safety standards that exist in the law, any psychological manipulation of the user would be set by industry self-policing. It is not just children that can be psychologically manipulated by AI as evidenced by how many of us have made impulse purchases based on suggestions provided by intelligent agents and again none of this is regulated, nor are the ethical concerns even thought about much by the companies that utilize this technology as a primary source of income.

These technologies when found in the business setting also include many of the same concerns as I have just raised but also include the growing use of smart malware to conduct theft and corporate espionage. If a machine can fool judges in a Turing test, one might repurpose this kind of software to more intelligently phish for passwords or sensitive information through extended interactions via text or email. Quite a lot of the customer service sector is now done through automation of various levels of sophistication but we have yet to have a coherent ethical discussion about what it means to take money from customers whom many companies largely refuses to deal with person to person.

Most of what has been said already applies equally as well to the field of medicine but the stakes are much higher as we are often talking about life and death decisions. An AI program that mistakenly denies the insurance claim of a patient may doom them to lesser care that could lead to a worsening illness or even death. The use of intelligent agents, typically in the form of robots to provide companionship to the elderly raises concerns regarding the ethics of fooling people into ascribing agency and care to a machine (Sharkey and Sharkey 2012b, 2010, Sharkey and Sharkey 2012a, 2011, Miller 2010). Additionally, robotic surgery has become a rapidly growing field and it has its own set of ethical concerns. Luckily, the medical industry has a history of taking ethical questions seriously and we are likely to see these issues addressed. For instance, David Luxton has noticed that, "[e]xisting ethics codes and practice guidelines do not presently consider the current or the future use of interactive artificial intelligent agents to assist and to potentially replace mental health care professionals" (2014, p. 1) and he has suggested some guidelines in a recent paper. Military and Police AI and robotics technologies also have an obvious set of ethical impacts, and indeed there is a

---

1 Affective computing is a process used to give the machine the ability to sense emotions in its users, model potential reactions, and simulate its own emotions in order to facilitate its interactions with humans.

large body of work already being done from the academic and political domains on addressing these issues. But an additional concern that is specific to our topic is that any company that accepts military AI and robotics contracts should have spent some time deliberating the ethics of doing so. They owe this to their investors and employees and should have a coherent justification for their decision to enter this industry. A further consideration is that a coherent argument can be made that an ethical organization should not be involved at all in the creation of autonomous and semiautonomous weapons systems, even if there is nothing illegal about working in this field (Sharkey 2010, 2012). An ethics board can help immensely in thinking through these difficult questions that might require avoiding a potentially lucrative market for ethical reasons.

Long Term Ethical Concerns

This is where our topic begins to sound much more like science fiction but we should not be too embarrassed by that. While accurate prediction of the future is impossible, we can still get close in our prognostications and intelligent things can be said about where we are going with these technologies in the distant futures and whether or not we really want to go there. One of the main values of good science fiction is that it can help us work through scenarios and possible ways the world might change in the future right here in the comfort of the present, and in so doing we might be able to avoid the dark worlds that we can imagine. There have been hundreds of depictions of the dystopian future that robotics and AI might bring. In fact the very word "robot" was coined in an early science fiction play that ended with the robots killing off their human masters. So we have a long history of being well aware of the danger that these technologies might pose to our survival and if we chose to ignore this advanced warning, then we would deserve our eventual fate. Recently, there has been some very public pronouncements by respected scientists and industrialists who warn of the danger that AI posses to the human race (Hawking et al. 2014) but it is unlikely that our future holds extinction at the hands of robot overlords, we have seen them coming for decades now and we can easily head them off by simply not inventing them.

There are some more subtle and interesting ethical concerns that will arise if we succeed at creating AI entities. The most profitable use for AI entities would be in the creation of AI/robotic companions. It is quite likely that many of the early adopters of expensive robotics technology might be motivated by the lure of acquiring a robotic companion or surrogate sex partner. One of the primary ethical concerns of this technology is that the companies that produce them could use the powerful motivation of sex to manipulate their customers through in app purchases or insuring brand loyalty through sexual favors (Sullins 2012). Evolution has preloaded us with psychologies that can be easily manipulated with sex and there are many more unethical ways to utilize this than there are ethical ones.

Even if these entities do not pander to our sexual desires, they may manipulate us in other ways economically. Children raised on AI toys may readily accept AI corporate mascots. One might actually develop a trusting relationship with a future fully robotic Ronald McDonald or Tony the Tiger through multiple conversations and economic interactions over the course of ones life.

Building on this, we have already set certain legal precedence in the USA in favor of the idea of corporate personhood that grants unconscious corporations with free speech rights. What other rights may follow? This legal environment might already be populated with many rights that AI entities might successfully sue for.

Once these entities have strong representation in the legal and political world, they may seek divergent goals for the use of this world that are different from our own. This potential situation is a lot more likely than the robot army we are so familiar with in science fiction. If we are not scrupulously careful at every step in developing these technologies, then given that we like material wealth and we like the idea of robot friends and partners, this might cause us to go too far down this path like lambs to the slaughter. Again, the time to make changes is the design of these technologies is now when these AI entities do not yet exist and therefore cannot effectively resist us.

Ethics boards of the more distant future will have the unenviable position of being one of the last locations where humans can significantly control these artificial entities. They will have to review the actions of these entities and ensure that they are consistent with the ethical vision of the organization and the society they inhabit. Given that these entities will be owned by the corporations that built them, that will give these corporations some leverage in ensuring a human standard of ethical behavior from these machines.

This brings us to the more complex topics of machine morality and roboethics, which is the study of how to properly program machines so that they can make intelligent decisions that are also ethically justified. There have already been some in-depth discussions of this topic (Anderson and Anderson 2011, Bostrom 2003, Gunkel 2012, Lin, Abney, and Bekey 2011, Wallach and Allen 2009, Hall 2012, Shulman, Jonsson, and Tarleton 2009, Yudkowsky 2001), and at least one proposal for an actual machine that could be used to govern the ethical decisions made by autonomous and semiautonomous machines on the battlefield (Arkin 2009).

While it is great fun to speculate about the future, it is important not to conflate short and long-term problems since the long term ethical issues are only science fiction. A good sober ethics committee firmly grounded in the present and very near future still has plenty to do in working on short term issues I introduced earlier.

The Purpose of AI and Robotics Ethics Boards

As we have seen in the last section, there is much pressing business that an AI and robotics ethics board can deliberate on. But we must also acknowledge that there are many ways to design and implement an ethics board and they are not all of equal value. A poorly designed ethics review process can sometimes seem worse than no overview at all. It would be useless to design a committee that served only a perfunctory role, perhaps only existing to allay investor worries but which was in reality too week to make any substantive changes in the design of the technologies under analysis. We could also error on the side of giving the board too much power. A closed and secretive committee that might set itself up as an obstacle to researchers and innovation will only engender fear and avoidance amongst the very constituency it is supposed to serve. These styles of ethics boards have existed in other disciplines of study and there is no need to recapitulate them in the field of AI and robotics research.

Ethics in a product design setting is best seen as the development of a dialog that engages all constituent groups involved in the design process. The invitation to these discussions must be given as widely as it is practical to do so to ensure that all those who might be impacted by the design have some input into its design. An ethics board should be the location that fosters that dialog both within the organization and with the wider customer base and public that the organization serves. Since engineers and programmers are the ones best suited to modify the project if potential troubles are discovered in this process, these engineers and programmers should be asked to contemplate the ethical challenges their new technology might occasion. It is also vital that these discussions happen early and are largely self-motivated by the researchers. A process where an ethics board inserts itself late in the design cycle and attempts to force some unwanted changes to a design from the outside will not be as successful.

This kind of open ethics dialog is a rarity in the world of compliance committees and IRB certifications. It has been recently reported that only ten percent of IRBs even invite the researchers to present their case to the board and that the researches often receive only a yes or no answer to their proposals with little to no feedback and rarely even know who was on the board that might have denied their research.

> Bureaucratically, ethics committee structures vary internationally and between institutions. Some IRBs are devolved, having the supervisor as one of the three- to four-person review committee. The majority of IRBs, however, meet behind closed doors. (Tolich and Tumilty 2014, p. 202)

One of the benefits of being a late addition to the world of ethics committees is that the AI and robotics world does not have to shackle itself with these old moribund ideas on the role of ethics in design. Ethics is not simply a set of standards to which one measures themselves to, it is instead an aspirational endeavor, a desire to honestly confront the likely outcomes of a proposed action. Ethical behavior is that which is guided by thorough and deliberative discussions about the situation that confronts us. It is difficult to pre-think all the contingencies involved in real-world moral problems, so there is no way a committee could be designed to simply rubberstamp projects that meet some arbitrary code of ethics and have that result be anything that is ultimately useful for the organization or the society it is a part of. Ethics is a territory that must be discovered and does not come to us all at once as if through revelation. The novelty of the situation at hand makes the discussion all the more necessary as it is hard for people to think through all the implications of technologies that do not yet exist. Given that AI and robotics applications are so new, we cannot rely on some set of ethical maxims derived in other intellectual domains and contexts, we must do the hard work of discovering them ourselves.

One alternative approach to the problem of developing conversations around developing technologies that looks promising is *The Ethics Application Repository* (TEAR), "TEAR sets out to break a cycle of fear and avoidance by facilitating better relationships between researchers and their IRBs" (ibid.). The TEAR approach allows us to develop a library of past lessons learned in dealing with specific case studies so that others can access that hard won knowledge for future work. An organization would do well to build a similar repository where researchers could explicitly explore the ethical impacts their proposed technologies might have by comparing their work to other similar projects and in addition this tool could be used to review the ethical impacts that technologies have had so far in each specific product's lifecycle. As issues arise, these documents could serve as a check to see if due diligence was observed and if solutions already exist. Technologies all have unintended consequences but sometimes those consequences could have been foreseen by more thorough analysis accomplished early in the design and implementation of a product. An additional benefit to this ethical research repository is that it allows those new to the field of technology ethics to have someplace to go to discover best practices which can facilitate the creation of communities of knowledge and expertise within the organization that could then specialize in resolving ethical problems specific to their own organization.

An AI and robotics ethics review board can meet the challenges of the short term ethical concerns by creating a dialog that follows these steps in its development.

1. Identify the ethical concerns raised by the new technology.
2. Anticipate consequences that may at this time be unforeseen by the product's designers. This creates a proactive ethics implementation rather than a reactive one.
3. Enhance the standard model IRB and replace it with one that fosters embedded ethicists in the design groups which closely work with the group to help foster a community of practice around ethical deliberation.
4. Vet the overall design strategy of the organization. Define the ethical goals of the company—what does the organization want to craft as its legacy?
5. Help operationalize the ethical code of the organization as it is applied to AI and robotic projects and update this code as new challenges are discovered.
6. Keep a repository of these deliberations to facilitate future discussions and manage the knowledge gained through the ethical review process.
7. Periodically reevaluate this process through a review of past cases. Officially celebrate the successes and work to not repeat failure.

This process will result in a deliberative practice where ethics is a respected part of the design process. With this method, designers themselves help identify potential ethical concerns and quickly work to mitigate them. This avoids the more common practice of not dealing with the issues until someone or some community is injured or annoyed by the product and the mess has to be cleaned up later. In the case of genuine unintended consequences that result in ethical problems, this process, at the very least, demonstrates documented due diligence and can serve as a way to expedite the design of a new solution given the record of what had been tried in earlier stages of the design or in other past projects.

Choosing the Right Ethics and Other Red Herrings

Some further recommendations are also in order. First, an ethics board is not the place to do the heavy lifting of metaethics, it is very safe to leave that pursuit to more philosophical settings. While I encourage every interested person to delve into the intricacies of ethical thought and argumentation for their own personal development, I don't think we are going to resolve any of the long standing deep philosophical debates in that field through the process I have outlined above. I agree that it would be easier if it were possible to resolve the question of which ethical system is best, and then apply the winner to all ethical problems we encounter but we are not going to be so fortunate. Since these are multigenerational discussions but we have pressing ethical problems that cannot wait for the theoreticians to settle their differences before we act. It is fine for now to work with the systems that have been shown to be valuable in computer ethics and other applied ethics contexts right now and add new ideas to the process as they are developed by professionals who work on these deeper issues. Luckily we do already have some powerful ideas that can be used today.

There are three main contenders in the field of metaethics that have stood the test of time and each has an important role to play in any applied computer ethics project. Consequentialism is the ethical theory that argues that the consequences of an action are what justify its ethical value. This idea has many applications but is best used to help fit ethics into an organization's business plan. Deontology, which is the idea that an action is right or wrong regardless of its consequences, has been successfully used to construct a company's founding values and in communicating its stance on upholding human rights and principles. Virtue ethics, which focuses on the development of good character in the individual, is best used to help guide the actions of individual researchers and designers who actually build and program the technologies in question. When it comes to AI and robotics we have an strange additional concern not fond in other applied ethics contexts and that is that if we are very successful, then we will have created new thinking beings and we will need to use all of the above ethical theories in the design of machine ethics systems to build an ethical character into the AI entities we create. Still, we must realize that all of these three ethical systems can be fundamentally at odds with one another at certain levels of analysis and that their use is part of a continuous academic dialogue which itself is contingent on the culture those conversations happen in. This means that all deliberations by an AI or robotics ethics board should be open and inviting of dialog and avoid getting trapped in the dogma that can be found by the most strident adherents of any one of these schools of thought. We must utilize the powerful ideas available from the large body of ethics research while realizing that their findings must be left open to appeal given that no universal ethical system has been developed yet that can defy all potential counter arguments against it.

Using and Creating Codes of Ethics

An ethics board can ease its job by building off of the documents that have been written already such as the ACM code of ethics (ACM n.d.), or the Accreditation Board for Engineering and Technology code of ethics (ABET n.d.). There are also numerous researchers in the fields of philosophy of technology and Philosophy and Computation that have already devoted full careers to the study of AI and Robotic ethics so there is already a body of research available for implementation.[2]

It is worth reiterating the strengths and weaknesses of developing written codes of ethics. They do nothing on their own, they must be reviewed and actually utilized by the employees as well as the management of an organization otherwise they have limited effect. It has been shown that even if one simply reads the ACM code of ethics, there is a short period where the reader will make better ethical decisions (Peslak 2007). Other studies have shown similar results which suggest that codes of ethics make an important contribution to the information and computing technologies (ICT) profession (den Bergh and Deschoolmeester 2010). So having researchers review the document as a team at certain intervals during their research could have at least a small positive effect. But ethics seen only this way is pedantic and boring and will not inspire the kind of ethical behavior we are after. There is no list of preconceived "thou shalt ...," propositions of any length that will be sufficient to guide our behavior in all but the simplest of ethical deliberations. In the end, ethical behavior is accomplished only by people already committed to its practice. Requiring ethical education of all of our programmers, researchers, and engineers is vital to making this work. The job of an ethics board is to help foster an environment where the ethical inclinations of people with good character can develop and flourish. An ethics board cannot be thought of as the only site for all ethical thought in the organization, absolving all others from ever having to think for themselves. Its duty is to start these conversations throughout the organization at every level.

---

2   See The Society of Philosophy and technology (SPT), International Association for Computers and Philosophy (IACAP), and the American Philosophical Association Committee on Computers and Philosophy (APA CAP), just to name three of the more prominent organizations in this area of research.

Finally, we must address the problem of the ethics board being seen as something of an outsider or some kind of referee that is not part of the team and is therefore an obstacle to be worked around. There will always be a little of that when people are brought into a team, especially when their expertise is not commonly encountered in the design process. It is also very natural for engineers and designers to hold preconceived resentment against ethics professionals if they have ever been the victim of some overzealous IRB process where their project suffered unjustly due to heavy handed bureaucratic overreach or incompetence or have heard stories from others that have suffered from this. We can eliminate most of the harmful effects of this hesitation to apply ethics by embedding ethicists as consultants into the research that an organization does and having them work as part of the team. In this way they can act as guides and early warning systems that can use their expertise to report to the ethics board on the developments that might be worth monitoring and in this way help researchers make better choices that can be justified to the media and politicians when questions arise from outside the organization. This is not to say that self-policing is all that is needed in ensuring ethical compliance. That would be a very foolish position to hold since sufficiently corrupt internal business practices might go to work on the embedded ethicist and they could become complicit in the process they were sent in to help fix. In egregious cases we can appreciate the important role that outsiders with the power to block unethical practices can play, this is where professional organizations, government regulators and nongovernment watchdog groups play a vital role in the demanding ethical behavior from those who are creating the world we all have to live in. Ideally an organization should never get to that point, but we all know that it has happened in the past and will continue to happen in the future but the above ideas will help the industry of AI and robotics to perform better than other industries might have in the past.

Conclusions

We have seen that it is time that AI and Robotics researchers seek some professional ethical guidance as they create some of the most potentially disruptive technologies in human history. It has also been shown here that ethics is best seen as the outcome of honest and open discussions within the design teams creating these technologies through a process that builds on the known best practices but which is also ready to change in light of the unforeseen consequences of deploying the technologies under development. This means it is important to create the right kind of ethics board and giving it the right amount of authority and autonomy without causing it to be seen as an outsider of obstacle to the goals of the team. Recreating the restrictive IRBs often found in other professions will not be useful to the design of AI and robotics technologies, but embedding ethical thought into the design of these machines will. Additionally, each organization must safeguard that the right mix of inside and outside influencers are on the board and embedded within the research teams themselves. This is critical to reduce the natural resistance that the board will receive from the organization at large and help it foster the best aspirations of the organization it serves while also allowing it to successfully question unethical practices within the organization that it discovers.

Even though AI and robotics is a new industry, it is not too soon to act. In fact, if organizations do not start paying some serious attention to the many ethical problems discussed in this chapter that AI and robotics research can create, then we may reach a point where it is too late to act and our worst fears may become realized. But we can be confident that this dystopia will not happen; we humans are too good at self-preservation and we have all the ethical tools we need now to secure our future.

References

ABET. n.d. "Neon Color Spreadingaccreditation Board for Engineering and Technology Code of Ethics." Accessed July 2014 https://www.iienet2.org/details.aspx?id=299.

ACM. n.d. "Task Force for the Revision of the ACM Code of Ethics and Professional Conduct, ACM Code of Ethics and Professional Conduct." Accessed March 2014 https://www.acm.org/about/code-of-ethics.

Anderson, M. and S.L. Anderson, eds. 2011. *Machine Ethics*. New York: Cambridge University Press.

Arkin, R.C. 2009. *Governing Lethal Behavior in Autonomous Robots*. Broken Sound Parkway, NW: Chapman and Hall/CRC.

Bostrom, N. 2003. "Ethical Issues in Advanced Artificial Intelligence." In *Cognitive, Emotive and Ethical Aspects of Decision Making in Humans and in Artificial Intelligence*, 12-17. Int. Institute of Advanced Studies in Systems Research and Cybernetics. Accessed July 2014. Retrieved from http://www.nickbostrom.com/ethics/ai.html.

Burrell, I. 2014. "Google Buys Uk Artificial Intelligence Start-up Deepmind for £400m." *The Independent*, January 27. Accessed July 2014. http://www.independent.co.uk/life-style/gadgets-and-tech/deepmind-google-buys-uk-artificial-intelligence-startup-for-242m-9087109.html.

Clark, H. 2006. "Chief Ethics Officers: Who Needs Them?" *Forbes*, October 23. Accessed July 2014 http://www.forbes.com/2006/10/23/leadership-ethics-hp-lead-govern-cx_hc_1023ethics.html.

Cleland, S. 2014. "Deepmind "Google Ethics Board" Is an Oxymoron, and a Warning – Part 11 Google Unethics Series." *PrecursorBlog*, January 30. Accessed July 2014. http://www.precursorblog.com/?q=content/deepmind-%E2%80%9Cgoogle-ethics-board%E2%80%9D-oxymoron-and-a-warning-%E2%80%93-part-11-google-unethics-series.

Cohen, R. 2014. "What's Driving Google's Obsession with Artificial Intelligence and Robots?" *Forbes*, January 28, 2014. Accessed July 2014 http://www.

forbes.com/sites/reuvencohen/2014/01/28/whats-driving-googles-obsession-with-artificial-intelligence-and-robots.

den Bergh, J. and D. Deschoolmeester. 2010. "Ethical Decision Making in ICT: Discussing the Impact of an Ethical Code of Conduct." *Communications of the IBIMA*:1-11. doi: http://dx.doi.org/10.5171/2010.127497.

Gannes, L. and J. Temple. 2014. "More on Deepmind: Ai Startup to Work Directly with Google's Search Team." *Re/Code*, January 27. Accessed July 2014. http://recode.net/2014/01/27/more-on-deepmind-ai-startup-to-work-directly-with-googles-search-team.

Garling, C. 2014. "As Artificial Intelligence Grows, So Do Ethical Concerns." *SF Gate*, January 31. Accessed July 2014. http://www.sfgate.com/technology/article/As-artificial-intelligence-grows-so-do-ethical-5194466.php.

Gunkel, D.J. 2012. *The Machine Question: Critical Perspectives on Ai, Robots, and Ethics*. Cambridge, MA: The MIT Press.

Hall, J.S. 2012. *Beyond Ai: Creating the Conscience of the Machine*. Amherst, NY: Prometheus Books.

Hawking, S., S. Russell, M. Tegmark, and F. Wilczek. 2014. "Stephen Hawking: 'Transcendence Looks at the Implications of Artificial Intelligence - but Are We Taking Ai Seriously Enough?'." *The Independent*, May 1. Accessed July 2014. http://www.independent.co.uk/news/science/stephen-hawking-transcendence-looks-at-the-implications-of-artificial-intelligence--but-are-we-taking-ai-seriously-enough-9313474.html.

Lin, P., K. Abney, and G.A. Bekey, eds. 2011. *Robot Ethics: The Ethical and Social Implications of Robotics*, edited by Ronald C. Arkin, George A. Bekey, Henrik I. Christensen, Edmund H. Durfee, David Kortenkamp, Michael Wooldridge and Yoshihiko Nakamura, *Intelligent Robotics and Autonomous Agents Series*. Cambridge, MA: The MIT Press.

Lin, P. and E. Selinger. 2014. "Inside Google's Mysterious Ethics Board." *Forbes*, February 3. Accessed July 2014. http://www.forbes.com/sites/privacynotice/2014/02/03/inside-googles-mysterious-ethics-board.

Luxton, D.D. 2014. "Recommendations for the Ethical Use and Design of Artificial Intelligent Care Providers." *Artificial Intelligence in Medicine* 62(1): 1-10. doi: http://dx.doi.org/10.1016/j.artmed.2014.06.004.

Matloff, N. 2014. "What's Wrong with Tech Leaders?" *CNN Opinion*, July 8. Accessed July 2014. http://www.cnn.com/2014/07/08/opinion/matloff-tech-ethics/index.html?hpt=hp_bn5.

Miller, K.W. 2010. "It's Not Nice to Fool Humans." *IT Professional* 12(1): 51-2. doi: http://dx.doi.org/10.1109/mitp.2010.32.

Peslak, A.R. 2007. "A Review of the Impact of ACM Code of Conduct on Information Technology Moral Judgment and Intent." *Journal of Computer Information Systems* 47(3): 1-10.

Prigg, M. 2014. "Google Sets up Artificial Intelligence Ethics Board to Curb the Rise of the Robots." *Daily Mail*, January 29. Accessed July 2014. http://

www.dailymail.co.uk/sciencetech/article-2548355/Google-sets-artificial-intelligence-ethics-board-curb-rise-robots.html - ixzz36RsuLtSv.

Sharkey, A. and N. Sharkey. 2011. "Children, the Elderly, and Interactive Robots: Anthropomorphism and Deception in Robot Care and Companionship." *IEEE Robotics & Automation Magazine* 18(1): 32-8. doi: http://dx.doi.org/10.1109/mra.2010.940151.

Sharkey, A. and N. Sharkey. 2012a. "Granny and the Robots: Ethical Issues in Robot Care for the Elderly." *Ethics and Information Technology* 14(1): 27-40. doi: http://dx.doi.org/10.1007/s10676-010-9234-6.

Sharkey, N. 2010. "Saying 'No!' to Lethal Autonomous Targeting." *Journal of Military Ethics* 9(4): 369-83. doi: http://dx.doi.org/10.1080/15027570.2010.537903.

Sharkey, N. 2012. "Automating Warfare: Lessons Learned from the Drones." *Journal of Law, Information & Science* 21(2). doi: http://dx.doi.org/10.5778/JLIS.2011.21.Sharkey.1.

Sharkey, N. and A. Sharkey. 2010. "The Crying Shame of Robot Nannies: An Ethical Appraisal." *Interaction Studies* 11(2): 161-90. doi: http://dx.doi.org/10.1075/is.11.2.01sha.

Sharkey, N. and A. Sharkey. 2012b. "The Eldercare Factory." *Gerontology* 58(3): 282-8. doi: http://dx.doi.org/10.1159/000329483.

Shulman, C., H. Jonsson, and N. Tarleton. 2009. "Machine Ethics and Superintelligence." In *Ap-Cap 2009: The Fifth Asia-Pacific Computing and Philosophy Conference. October 1st-2nd*, edited by Carson Reynolds and Alvaro Cassinelli, 95-7. University of Tokyo, Japan.

Sullins, J.P. 2004. "Artificial Intelligence." In *Encyclopedia of Science, Technology, and Ethics*, edited by Carl Mitcham. 1st edtion, 2nd edition in press. US: Macmillan Reference.

Sullins, J.P. 2006. "When Is a Robot a Moral Agent?" *International Review of Information Ethics* 6: 26-30.

Sullins, J.P. 2011. "Introduction: Open Questions in Roboethics." *Philosophy & Technology* 24(3): 233-8. doi: http://dx.doi.org/10.1007/s13347-011-0043-6.

Sullins, J.P. 2012. "Robots, Love, and Sex: The Ethics of Building a Love Machine." *IEEE Transactions on Affective Computing* 3(4): 398-409. doi: http://dx.doi.org/10.1109/t-affc.2012.31.

Sullins, J.P. 2014. "Information Technology and Moral Values." In *The Stanford Encyclopedia of Philosophy*, ed Edward N. Zalta. Spring 2014 Edition. http://plato.stanford.edu/archives/spr2014/entries/it-moral-values (accessed July 2014).

Tolich, M. and E. Tumilty. 2014. "Making Ethics Review a Learning Institution: The Ethics Application Repository Proof of Concept - tear.otago.ac.nz." *Qualitative Research* 14(2): 201-12. doi: http://dx.doi.org/10.1177/1468794112468476.

Wallach, W. and C. Allen. 2009. *Moral Machines: Teaching Robots Right from Wrong*. Oxford: Oxford University Press.

Weaver, J.F. 2014. "What a Supreme Court Case Means for Google's and Facebook's Use of Artificial Intelligence." *Slate.com*, February 3. Accessed July 2014. http://www.slate.com/blogs/future_tense/2014/02/03/deepmind_ google_ai_ethics_board_what_u_s_v_jones_means_for_tech_companies. html.

Yudkowsky, E. 2001. "Creating Friendly Ai 1.0: The Analysis and Design of Benevolent Goal Architectures." San Francisco, CA: The Singularity Institute. http://intelligence.org/files/CFAI.pdf (accessed July 2014).

Chapter 6

Technological Dangers and the Potential of Human–Robot Interaction: A Philosophical Investigation of Fundamental Epistemological Mechanisms of Discrimination

Marco Nørskov[1,2]

The ethical debate on social robotics has become one of the cutting edge topics of our time. When it comes to both academic and non-academic debates, the methodological framework is, with few exceptions, typically and tacitly grounded in an us-versus-them perspective. It is as though we were watching a soccer game in which our child is one of the players. The question of which team we should cheer for never occurs to the parent. By changing the vantage point to a radical phenomenological perspective, informed by Eastern as well as Western thought, this chapter tests the basis for this type of positioning with regard to HRI. It is argued that the process itself is an artifact with moral significance, and consequently tantamount to discrimination. Furthermore, influenced by Heidegger's warnings concerning technology, this chapter explores the possibilities of HRI with respect to the accompanying technological dangers and opportunities. Finally, aiming for the very limits of the theory, I discuss the contours of a praxis facilitating *being-with-robots* beyond conceptualization. Basically, this mode of being, pertaining to

1   An earlier version of this work appeared as Nørskov, M. 2014. "Human–Robot Interaction and Human Self-Realization: Reflections on the Epistemology of Discrimination." In *Sociable Robots and the Future of Social Relations: Proceedings of Robo-Philosophy 2014*, edited by Johanna Seibt, Raul Hakli and Marco Nørskov, 319-327. Amsterdam: IOS Press Ebooks. doi: http://dx.doi.org/10.3233/978-1-61499-480-0-319. Republished with permission of IOS Press.

2   I am deeply indebted to my colleague, Johanna Seibt (Aarhus University), who commented on the chapter throughout its various production stages, and from whom I learned so much. I would also like to thank Kohji Ishihara (The University of Tokyo) for his most helpful critical and constructive comments on an earlier draft of this chapter. This work was partially supported by the VELUX Foundation under the PENSOR Project.

Social Robots

non-technological HRI, bypasses Heidegger's warnings, and potentially facilitates a certain kind of self-realization for the human involved.

Introduction

Robotics is one of the cutting edge research areas of our time, and it has become a truly interdisciplinary field, integrating various disciplines, such as engineering, computer science, psychology, philosophy, etc., into its domain. Commercial, military applications, as well as public robotics research projects, attract massive funding and constant media attention. This should be no great surprise, as robots are often heralded as promising solutions to many of our contemporary problems, for example, demographic challenges and ethical warfare. They are not only supposed to free us from the burden of the so-called "3Ds," that is, "dirty," "dangerous" and "dull," or jobs, but to provide us with novel types of interaction partners, *sociable robots*, who/which are integrated into our social sphere (cf. Takayama, Ju, and Nass 2008, p. 25). Nevertheless, robotics not only infuses the dreams of the utopian with fresh hope, it also causes nightmares to the dystopian thinker. The prospect of robotic applications being restricted by our fantasies and technical limitations only, naturally begs a broad range of ethical questions. Is it morally right to outsource caretaking to robots (e.g. Sparrow and Sparrow 2006)? Are androids the only responsible and culpable agents (Nadeau 2006)? Do robots deserve ethical concern as patients or agents (cf. Gunkel 2012)? These are just a few of the various ethical concerns emerging in connection with the outlook on the development of these novel machines. The selected examples also highlight the fact that current ethical inquiries are not only restricted to our moral obligations towards *Homo sapiens sapiens*, but also reflect the possibility and adequateness of ethical obligations towards various types of robots. Finding answers to these issues is no easy undertaking, as the capacities and features of robots vary, and are constantly enhanced. As the quote below indicates, we have a duty to apprehend robots correctly.

> [… F]ailure to apprehend the world accurately is itself a (minor) moral failure. We have a duty to see the world as it is. It is a sad thing to be deceived about the world; it is a bad thing to perpetuate and prolong such deception ourselves. (Sparrow and Sparrow 2006, p. 155)

As pointed out in the literature (Gunkel 2012), on a fundamental level, it is we who conventionalize who and what is to be included into the circle of beings deserving *ethical consideration*. This chapter at hand follows this line of thought, and aims to provide philosophical insight into the epistemological "mechanics" behind this process. By merging the discourse with the realm of Buddhist-inspired thought, where phenomena simultaneously withdraw and disclose themselves from conceptual determination, I analyze the amalgamation of epistemology and its moral significance in the course of apperception. From this radical phenomenological standpoint (Section: *A Contrasting Framework for Inspiration*), it may be argued that the "as if" (cf. Turkle 2012, p. 54), that is, the robotic simulation of something a robot is not (e.g. a cat, a human), is an artificial construct on a fundamental level, and furthermore, owing to the moral significance of the process, the distinction between the robot and what it simulates pertains to concept of discrimination (Section: *Application to HRI*). As a consequence of this process, it seems only natural to regard robots exclusively as lifeless servants or, at Heidegger's level of abstraction, as *standing reserves*. Not being aware of this process presents a rather serious danger, but also fosters n the potential for realizing our dignity, as suggested in the *Question Concerning Technology* (Heidegger 2004). Without subscribing to any form of orthodoxy, the Buddhist-inspired thought will be used to unfold this potential, as well as the dangers, in the context of *Human–Robot Interaction* ((HRI) Section: *Dangers and Potential*), and maybe more importantly, outline the possibility of a praxis involving robots that aims for human self-realization as γνῶθι σεαυτόν, and the cultivation thereof (Section: *HRI and Human Self-Realization*). Finally, the chapter is rounded off with some concluding remarks.

Let me first offer a brief outlook on the current debate on social robotics, in order to provide an insight into what is at stake, the rationale behind it, and the limitations of traditional inquiries.

A Glimpse into the Public Ethical Debate

Before entering into the phenomenological examination, it is fruitful to take a closer look at the current public ethical debate, in order to ground our considerations in a real-world context. In addition, a few selected examples of concrete inquiries by policymakers and public discourse on HRI will draw attention to the particular way in which the various ethical concerns are framed.[3]

Ethical and moral considerations pertaining to robots used to be mainly addressed in science fiction novels. But in recent years, they have received increasing attention, and are taken up as serious subjects by various debaters, ranging from professional philosophers to the media. Also, the issues have not escaped the attention of various public organizations and institutions. For example, in November 2006, the government of South Korea initiated work on a *Robot Ethics Charter* (Lovgren 2007), and in Denmark, the so-called *Danish Council of Ethics* recently published guidelines with respect to service robots (Det Etiske Råd 2010c) and cyborgs (2010a).

---

3   Several of the here presented thoughts have been considered to some extent in my PhD dissertation (2011) and have been substantially revised and matured for this Chapter.

In 2006, the *Italian EURON Roboethics Atelier* offered the following list of problematic issues pertaining to the implementation of robotics in human environments:

> We have forecasted problems connected to:
>
> 1. *Replacement of human beings* (economic problems; human unemployment; reliability; dependability; and so on)
>
> 2. *Psychological* problems (deviations in human emotions, problems of attachment, disorganization in children, fears, panic, confusion between real and artificial, feeling of subordination towards robots).
>
> 3. Well before evolving to become conscious agents, humanoids can be an extraordinary *tool used to control human beings*.[4] (Veruggio 2006, p. 617)

Here, this list is supplemented by a fourth concern, concretized in a text by the *Danish Council of Ethics*:

> [O]ne can ask to what extent it is ethically problematical in itself, if social robots become increasingly like humans in appearance, communication and behaviour and thereby pretend that they are independent, sentient active beings, i.e. 'as if' they are human beings. (Det Etiske Råd 2010b)

Even though to some extent this is an instance of the second issue, it is sufficiently central to be highlighted, and thus, I would like to add it under the following label:

> 4. *Simulation*[5] (e.g. misapprehension)

For the concerns of this chapter, issues 1, 2, and 4 are primarily important, so let us look at some concrete elaborations of these.

The first issue, *replacement*, currently has a particularly striking illustration in the Danish debate about service robots in the healthcare and geriatric sectors, with fears and expectations running unusually high—see, for instance, the online newspaper article, *The Danes are afraid of robots*[6] (Politiken 2008). Politicians who promote robot technology find themselves in a situation where they need to continuously reassure the public that the resource savings that new technologies

---

4   Numeration and italics supplied.

5   Addressing the vagueness with respect to attribution of capacities to robots in current research, Seibt recently (2014) introduced the five concepts of simulation, based on process-ontology, i.e. functional replication, imitation, mimicking, displaying, and approximation. In this chapter, these terms are used in this sense.

6   Translated by the author—the original title: "Danskerne er bange for robotter"

could realize would not result in downsizing of employees, but in a mutually beneficial redistribution of resources.[7] The argument is that by introducing robots for menial tasks, such as cleaning, the elderly would get qualified human care in other areas, for example, their psychological needs. The fear of replacement by robots might be culturally conditioned, however. Kitano (2005) claims that the ideas of robots taking over the world or stealing jobs from people are rooted in Western culture, and were originally not to be found in a Japanese context.

The concern about the *psychological effects* of introducing robots into society, issue 2, above, has a striking illustration in a recent documentary on service robots by Ambo (2007). The documentary follows several persons, including K,[8] an elderly resident of a nursing home in Germany. Since K has no visitors to entertain her, she received PARO as her companion, a seal robot invented by Takanori Shibata. In the narrative of the documentary K, seems to treat the machine like an animal, or even a child. Many of K's co-residents seem very puzzled by K's close relationship to the machine. This naturally raises the question of whether human beings devaluate themselves in their interactions with modern technology: is K subjecting herself to the mandate of the robot?

When it comes to *simulation* (issue 4), it seems to be tightly connected to the human capacity to perceive inanimate objects *anthropomorphically* or *zoomorphically*, as animate. Famous robots such as PARO, the seal robot, and KISMET, a robot in the shape of a head, are just some of the robots mentioned in the literature as examples of lifeless entities to which its users attribute animation, to a certain extent (cf. Turkle 2012). In this context, *The Danish Council of Ethics* raises two potential concerns (Det Etiske Råd 2010c: 12-14). First, they worry about the consequences of the intended or unintended "deceit," and second, the Council is concerned about the potential transformative power of HRI on "genuine" human interaction, which could lead to narcissism and egoism. These and similar warnings are also issued by various researchers, such as Sparrow and Sparrow (2006) or Turkle (2012).

For the further purposes of this chapter, it is not only important to acknowledge the social relevance and emotional depth of the metaphysical and ethical issues surfacing in the debate on HRI, but also to draw attention to the way in which it has framed its issues, that is, the basic assumptions that underlie the very foundation of the debate. As the few examples have already illustrated, the debate about robots basically wrestles with the nature of Western ethics, which is primarily concerned with providing us with norms for interpersonal interaction and the identification of ethical values that should guide human life. However, it seems difficult to determine the interactive capabilities of robots that would include them in the community of moral agents. To be sure, Western ethicists occasionally discuss the collective and individual ethical obligations of humans towards nonhuman

---

7   For instance, see the position of the Danish Minister of Social Affairs with respect to robotics in the geriatric sector, in the June issue of *Ældre Sagen NYT* (DaneAge 2009, p. 3).

8   The full name is not relevant to the chapter, and hence omitted here.

entities, such as animals or nature, but these are marginal developments. What fuels the concerns above, is, on the one hand, (1) a static concept of what a robot is, and (2) the basic assumption that human beings are something special. With respect to (1), there are at least three serious problems: The term "robot" covers so many different types of machines, and is understood in so many different ways that it seems to include almost everything and nothing at the same time. Furthermore, the sphere of robots is constantly expanded by novel inventions. Even with respect to the analysis of the functionality of certain robots, conceptual determination is pushed to its very limits (cf. Nørskov 2014b). Returning to the question of humans being special (2), it seems that what counts as "properly human" is also a rather dynamic concept (Gunkel 2012), as mentioned in the introduction, and it would simply be nice to have a stronger argument than what boils down to " …, because we are special," as this basically gives us carte blanche to do whatever we want.

The preconception that human interaction has a higher value than HRI seems to be generally tacitly accepted, and is the vantage point of traditional analysis. In short, our illustrations from the ethical debate about the social employment of robots conveys a bottleneck situation, where the theoretical tools available do not seem to fit the task to which they are to be applied. As I suggest in the following sections, one way of constructively addressing this problem is by moving beyond the fundamental assumptions of the standard theoretical framework, and this is most easily done if we expose ourselves to another tradition of philosophical thought. This does not necessarily imply a suspension of existing theories or moral imperatives as such, but the aim is to explore the qualitative potential of HRI in new ways. By applying Buddhist-inspired thought, in the next sections I investigate whether or not the discrimination between humans and robots is tangible, from a radical, phenomenological viewpoint.

A Contrasting Framework for Inspiration

As outlined above, in general, *roboethics* seems to be chained to and by the assumption that humans are animated by certain capacities, whereas robots are incapable of meeting these standards, and thus are fundamentally different. In this framework, it is natural to ask questions such as, to what extent and under what conditions do robots belong to the class of Cartesian or Kantian subjects, that is, of moral agents that are self-conscious, free, and rational? However, I pursue a different approach here, which aims to be a rather fundamental theoretical revision. Below, I suggest that some of the ethical implications of HRI reveal themselves, if we understand such interactions in a new guise, which may be called *being-with-robots*, resonating with Heidegger's terminology. In order to develop such a re-categorization of the phenomenon of HRI, it will prove useful to begin with a closer look at a philosophy that differs radically from the Cartesian/Kantian conception of the subject, namely, the rich thinking that may be found within Zen Buddhism, which is said to generally discard the existence of a permanent self, as such (Smart 1997, p. 309). I begin by introducing some key concepts of Buddhism, the notion of *suffering* and *bodymind awareness*, before turning to a discussion of HRI from a Buddhist-inspired point of view, in the next sections.

The turn to Buddhist philosophy may strike one as an unnecessary, meta-philosophical extravaganza to highlight fundamental presuppositions of the debate about robot ethics in Western countries. It is important to note (1) that it is a philosophical tradition in its own right, with proponents such as Dōgen, Nishida, and Nishitani, and furthermore, (2) that Western and Eastern societies are often described as reacting rather differently to the prospect of increased HRI (e.g. with respect to ethics, see Kitano 2005). This will suffice as motivation for a closer look at Eastern thought, especially in connection with HRI.

Suffering

Siddhartha Gautama, the historical Buddha, is commonly regarded as the worldly founder of Buddhism. The exact dates of his birth and death are disputed, but he was probably a near contemporary with Socrates (Cousins 1998, Smart 1997). Confronted with the hardship of the life of his father's vassals, so the story goes, he was determined to find a way to stop suffering (*ku*). His ambitious quest resulted in the rich and unique tradition of Buddhism, in its manifold varieties that developed during its historical spread from India to China, further to Japan, and to the rest of the world. One of the first core teachings of Buddhism attributed to the historical Buddha himself is *shitai* (Four Noble Truths), listed below (Mejor 1998):

- *Kutai* (There is suffering)[9]
- *Jittai* (There is a cause of the arising of suffering)
- *Mettai* (There is an extinction of suffering)
- *Dōtai* (There is a means to come to the end of suffering).

It is hardly deniable that there exists some sort of suffering in the world, certainly in the sense of the classical sufferings referred to as *kutai*, which are *birth, old age, illness*, and *death* (Akira 1990, pp. 39). However, in order to understand its depths in this context, it is necessary examine it further. With respect to the use of the word "suffering," Abe points out (probably already with the second truth in mind):

> The more we try to cling to pleasure and avoid suffering, the more entangled we become in the duality of pleasure and suffering. It is this whole process which constitutes Suffering [capitalized in the original]. When Gautama the Buddha says 'existence is (characterized by) suffering', he is referring to this Suffering and not to suffering as opposed to pleasure. (Abe 1985, p. 206)

---

9   The bracketed text following the names of the respective truths are not literal translations of the terms, but short descriptions of the meanings of these terms, which I have adapted from the Routledge Encyclopedia of Philosophy (Mejor 1998).

In this senses it is not so much the manifestation of what we value as "good" or "bad" that constitutes suffering; rather, as Abe suggests (ibid., p. 205), both are interdependent and it is the acting human disposition, distinguishing between the two, and preferring one to the other, that is suffering. This naturally brings us to the second truth, namely *jittai*, which explicitly describes the origin of suffering, and motivates much of the analysis in the following section. The origin of suffering, in the sense above, is craving for attachment in all its shapes, such as the desire to possess a sports car, or to be famous, but an obsessive nihilist's self-destructive quest would also be problematic. The reason for this is not so much a psychological but an epistemological issue, as is apparent from the following quote by Kasulis.

> ... Zen Buddhism criticizes our ordinary, unenlightened existence by refusing to accept a retrospective reconstruction of reality. (Kasulis 1985, p. 60)

Hence, by interpreting experience in terms of a fixed conceptual framework or categorization, we somehow distort the immediacy of perception as it occurs during our interactions with another entity, for example, a robot. Simply put, our way of conceptualizing does an injustice to the concrete phenomenon of experience, and as a consequence, deprives us of being in the moment with the phenomenon as it appears without conceptual prejudice. To sum up, the *suffering* focused on here is the danger of being anywhere but in the here and now, consequently influencing our quality of life, literally affecting our very presence.

Now that I have outlined what is at stake, I will introduce a description of experiential modes, which, as theoretical hermeneutics, will prove helpful with respect to analyzing the nature of HRI.

Bodymind Awareness

At what point do we discriminate between the phenomena of our perception, of, say, a robot and a human being? Buddhist philosophy has a lot to offer in this regard, as the mode of experience by which we perceive the world and its manifold phenomena is scrutinized to the borders of rationality and beyond. Therefore, in order to work out some of the insights from the previous subsection in greater detail, I use Shaner's classification of experience from his book, *The Bodymind Experience in Japanese Buddhism* (1985), as a vantage point for further investigation.

In his analysis of the *bodymind* in the records of Kūkai (774-835) and Dōgen (1200-1253), two of Japan's most influential religious figures and intellectuals, Shaner uses the realm and terminology of Husserl's phenomenology as a fundamental narrating tool.

> Mind-aspects and body-aspects may be separated from the context of lived experience only through abstraction. It is only through abstraction that mind-aspects or body-aspects may become noematic foci, for example, in

reflection, imagination, or reverie. In daily life we act as though the meaning of our experience simultaneously includes both mind-aspects and body-aspects. Accordingly, henceforth we will refer to the presence of both aspects in all experience as 'bodymind.' (Shaner 1985, p. 45)

Shaner uses the term *bodymind* to stress the inseparability of the mental and somatic in the very instance of experiencing. The oneness of body and mind is termed *shinjin ichi nyo* in the Buddhist tradition (Kasulis 1985, pp. 90-91). According to this theory, we would be mistaken in isolating the act of experiencing exclusively to the mind or the body. As a consequence, the traditional Cartesian distinction of mind and body into externally related units is something that happens only retrospectively, after the experience. It is a way of making sense of the world within our perceptual and conceptual limits, as we as individuals are not able to intellectually access and grasp all the phenomena and causal relations of reality at the same time. This observation resonates strongly with results in cognitive science (cf. Frith 2007, Hood 2013).

Returning to the *bodymind*, Shaner introduces several modes of what he terms *Bodymind Awareness* (henceforth: BA). The central concept in Shaner's classification of modes of BA is the notion of a "noetic vector," an application of Husserlian terminology. Shaner translates a *noetic vector* simply as "one's attention" (Shaner 1985, p. 52). However, and in agreement with Shaner's use elsewhere in the book, in a mathematical sense a vector is something that has direction and length, and a *noetic vector* may *eo ipso* be understood as the directedness of awareness. The difference between the three modes of BA is the number of *noetic vectors* involved at a given moment. The following table provides a brief overview, which should be sufficient for our purposes (for details ibid., p. 48). Additionally, it contains some of Shaner's examples, which help to relate the modes to concrete experience.

Table 6.1   Orders of Bodymind Awareness

Order of BA	Characteristics	Example
1st order BA	Here we do not force any noetic vector by intention but let the experiences come and go	Pre-reflective consciousness in which there is no intentionality
2nd order BA	A state with only one noetic vector	An athlete during intense competition (where he is utterly focused on the objective, e.g. hitting the target)
3rd order BA	Multiple noetic vectors	Reflective discursive consciousness

For the rest of the chapter I will denote these three different orders as BA1, BA2, and BA3 respectively, where BA1 corresponds to the *1st order BA*, etc.

Shaner's point is that given that we can never be sure of having grasped the world as it really is, BA1 is what lets us come the closest to the experience of it. As soon as we have one *noetic vector*, which usually occurs when, for example, we describe things, we have already altered the experience by narration. This process could be compared to Heisenberg's *Uncertainty Principle* (1927), or the "observer effect" in psychology, and is even more likely and more pronounced in BA3.

Building on Shaner's classification, and returning to our initial exposition of suffering, it is within BA2 and BA3 that suffering, as outlined in the previous subsection, occurs. A *noetic vector* may be interpreted as craving attachment, since it represents a fixation. It is important to note that there is nothing particularly wrong with limiting one's perception to one or a finite number of *noemata*, for instance, deliberately focusing on something, or connecting the various *noemata* in the process of reasoning, as such. However, it is equally important to note that in the process of narrowing the picture theoretically, to a certain focus (represented by the limited number of *noetic vectors* and the fixed standpoint from where they emerge), the rest of the picture will, nevertheless, always escape our observation.

Consequently, perception beyond BA1 is a process of selection and distortion, and it is on this fundamental epistemological level that I will explore the consequences of discrimination with respect to the human–robot encounter.

Application to HRI

Let us now apply the previous considerations to HRI, focus on the case of simulating robots, and investigate the soundness of the distinction between these robots and what they simulate.

At first blush, it seems that the issue may be quickly dealt with, since robots, in the current state of the arts, are clearly limited in their functionality, and cannot fully, *functionally replicate* the behavior of a human being, for instance, and therefore, the distinction seems perfectly well grounded. But let us take a closer look at the issue of simulation itself. Service robots are being built, some to resemble humans, such as Ishiguro's androids (Ishiguro 2006), others deliberately designed to look like something unfamiliar, as this seems to strengthen the robots' odds of not being regarded as simulating something they are not. For example, PARO, the famous robot seal, is deliberately designed as a seal because few people have been in close contact with a such an animal (Marti et al. 2005).

> Shibata [the inventor of PARO] says he tried making robotic cats and dogs, but that people didn't find those convincing. 'They expected too much,' he says, and would compare the robot to real animals they had known.

> Few people have ever seen a live baby seal, so they aren't likely to draw comparisons between the robot and the real thing. So they accept Paro as a cute little companion. (Greenfieldboyce 2008)

But why should we at all be concerned with the degree to which a robot, in appearance or function, simulates a living being, for example, a cat robot that is not a cat?

For Sparrow and Sparrow (2006) the issue of deceit is central, when it comes to social robots in the healthcare sector. Manifesting the appearance of having genuine capacities, such as emotions, may be harmful to the vulnerable elderly who have been "tricked" (e.g. K, noted above).

> Instead of being positive experiences which improve the lives of those who have them, 'affection' or 'love' for, or pleasure in the 'company' of, a robot pet are sentimental excesses that add nothing to a human life. The ability of sophisticated robots to provoke such emotions is not a virtue; it is a danger. (Sparrow 2002, p. 315)

> Insofar as robots can make people happier only when they are deceived about the robots' real nature, robots do not offer real improvements to people's well-being; in fact the use of robots can be properly said to harm them. (Sparrow and Sparrow 2006, p. 155)

It is debatable whether or not it is really sentimentality we are dealing with here (Rodogno 2014). Furthermore, taking a different ethical vantage point changes the outlook. For instance, Sharkey and Wood (2014) seem to come to a different conclusion in their utilitarian, cost-effectiveness analysis of PARO. Taking Sparrow and Sparrow's and others' criticism seriously, they nevertheless conclude:

> At the present moment, the likely benefits could be seen as justification for the risks. In the future, if the balance of the evidence tips more decisively towards the positive benefits of the Paro, it could even be argued that they *should* be made available to those people with dementia likely to benefit from them. (ibid., p. 4)

It should be rather clear from the short detour into the selected literature that social robotics is "mission critical." In the following paragraphs I would like approach the issue of robotic simulation from a different angle.

On the basis of the Buddhist-inspired epistemology just sketched, a philosophical explanation of an encounter with a robot could run as follows. In BA2 and BA3 our intentionality is actively put in relation to our previous experiences. BA2 and BA3 operate with fixed content that allows for comparison between the *noemata* of the present intention and previous experience. Hence, if we relate to our surroundings at the level of awareness of BA2 or BA3, we may, by having a categorical reference point, recognize when a robot is simulating something it is not, for instance, a cat or a human being. Given the current state of the art, robots are immediately perceived as not living up to our construct of what a cat should be like, for example, based on our experience. Accordingly, we become aware of the misfit, and this information influences how we think about, and ultimately, how we treat the robot.

To some extent, the just sketched theory also seems to be rather compatible with Mori's famous *The Uncanny Valley* (bukimi no tani) hypothesis (Mori, MacDorman, and Kageki 2012). Based on his experience, Mori plotted *affinity* and *human likeness* of robots against each other. The resulting graph shows that the more human robots become, the more *affinity* do we attribute to them. However, there is one exception. From a certain threshold, the function reaches its first peak, and thereafter, the *affinity* decreases, and even becomes negative. This is the *uncanny valley*, where we find zombie robots and the like—machines whose company makes us uncomfortable, and induces an eerie feeling. Nevertheless, the function of the graph again increases, exponentially, after the *uncanny valley* is passed. Here, robots become extremely humanlike, resulting in the second peak. Mori suggested in his original article of 1970 that robot engineers should aim for the first peak, as this would make it more likely that they would end up with a product that is appealing to its users. The theory above may interpret the shifts with respect to *affinity*, in Mori's theory. As long as a robot is not simulating a human too well, deliberately or not, we do not need to match it against our concept of what it means to be a human. Hence, in this regard we are not tempted/afforded to emerge into BA2 or BA3 to figure out whether or not we are dealing with a human being, and can simply accept the robot as it is. As soon as we enter the grey area, where we are not sure of anymore of what we are dealing with (e.g. is it a human being, does it have consciousness?), we find ourselves in BA2 or BA3, where we try to contrast it against our concept of "human being." Moving further along the *human-likeness* axis, the robot moves beyond the *uncanny valley*, where we find ourselves in BA1 again, with respect to the categorical matching. Here, we do not question the robot's human qualities, as we cannot, and hence do not discriminate, as there is no reference point.

Our sketch of Buddhist epistemology may be applied more generally. In fact, if one accepts the idea of levels of BA, any conceptualized experience is actually already *discrimination* in its truest sense.

> Even if all our senses are intact and our brain is functioning normally we do not have direct access to the physical world. It may fell as if we have direct access, but this is an illusion created by our brain. (Frith 2007, p. 42)

Given epistemological limitations with respect to accessing the physical world as outlined in the quote above, conceptualization becomes more than mere *differentiation*, *delimitation*, or *demarcation*, since it manifests as a sort of prejudice grounded in the inference from our previous conceptualizations to the present case. This prejudice happens as soon as we leave BA1, and we cannot avoid it, if we want to avail ourselves of the cognitive capacities that conceptualization provides, such as planning, imagining, and learning. Any intentional experience at the level of BA2 or BA3, whether of a human being or an object, immediately induces a classification, however retrospective, of what is experienced, for example, animate or inanimate. Importantly, both the experiencing subject and experienced object are products of perception, constructed by the determining process of perception. Nishida—according to Abe the "most outstanding philosopher of modern Japan" (Abe 2003, p. 66)—formulates this insight most radically:

> It is not that there is experience because there is an individual, but that there is an individual because there is experience. (Nishida 1990, p. 19)

Going by the common notion of "discrimination," the differentiation that characterizes discrimination is expressed in our ability to objectify or, in this context, in our ability to dehumanize, which is sketched in Figure 6.1 below as the transition from classification under the category of "Human" (H), to classification under the category, "nonhuman Object" (O). This cognitive transition is often cited in descriptions of slavery or the treatment of animals as commodities (O) (e.g. in Regan 2005). If we consider modern robotics, such as Ishiguro's androids, which aim for the second peak in Mori's *uncanny valley* theory, we can see a trend with respect to robot design that goes the other way—robots are intentionally designed to facilitate the association with human features. The robot as an object (O') is deliberately humanized to quasi-human (H') status, converging towards the status of a human being.

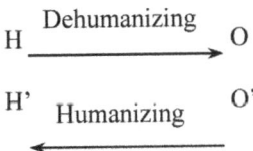

Figure 6.1   Objectifying and Humanizing

The cognitive transition of *dehumanization* (H→O) is something we normally consider ethically condemnable. However, with respect to robots, we are now in a situation where the inverse cognitive transition of *humanization* (O→H) raises concerns, too (cf. issue 4 in the previous section).

Figure 6.1 presents only a small part of the larger epistemological picture worked out in Buddhist philosophy. I am using it here not only to illustrate degrees and directions of conceptual superimposition, but also to highlight the link between conceptualization and moral valuation, which are not as separate in

Buddhist thought as the Western disciplines of Epistemology versus Ethics make out. In reviewing Dōgen's position[10] on ethics based on *Shōbōgenzō* (Treasury of the Correct Dharma-Eye) Kasulis writes:

> [... I]f one produces no categories to superimpose onto prereflective experience, there can be no evils at all. ... Since, following Nāgārjuna's argument, terms like good and evil are interdependent concepts, they operate on the level of thinking, but nonproduction is prior to such categorizations. (Kasulis 1985, p. 95)

> Only if the ego has been extirpated can there be compassion without distinctions and wisdom without presuppositions. (ibid., p. 98)

BA1 is equivalent to what Kasulis terms pre-reflective experience or the extirpated ego in the two quotes above. From the phenomenological perspective, it is when we first leave BA1 that the duality of good and evil arise. The mode of BA1 is beyond this distinction. Before continuing, it is important to note that this is different from asserting that we cannot engage in unethical activities while in BA1. An action itself may be evaluated against ethical standards by others, or in retrospect. Nevertheless, the subtle state of BA1 and the shift towards the other modes is the central topic of this investigation.

Going back to the example above, when we encounter a cat-like robot, we try to conceptualize the entity as a cat, but this process fails, somehow. The robot-cat does not qualify as a real cat, as it is not truly recognizable as such, based on pervious experience. Becoming aware of an entity coming close to being a cat, but ultimately failing to pass the threshold induces a subtle effect. Our immediate reaction is to associate the entity with not meeting the standards of the superimposed category, instead of viewing our brain as a central stakeholder in this evaluation process. It is important to remember that, as Frith states in his book, *Making up the Mind* (2007), we are not perceiving the world but our brain's model hereof (ibid., p. 132). We are wired to function in this way, however it is important to be aware of the partiality of this mechanism, as I outline in the remainder of this section. Our perception of the robot-cat amounts not only to a differentiation, but ultimately also to *discrimination*, as the combination of an ontological and an epistemological judgment bears moral significance. This may not be intuitive at first, and needs further explanation. As outlined above, we cannot claim complete access to the world in BA2 and BA3, or to reference cognitive psychology, we only approximate the world (cf. Frith 2007, Hood 2013), so depriving the cat-robot of moral standing, or indifference to this question as a consequence of category ascription (it not being recognized as a cat) has a moral significance, as it affects my relationship to it. This does not necessarily mean that

---

10   According to Bein, it was Watsuji who brought Dōgen's thought into the realm of modern philosophical inquiry (Bein 2011, p. 1).

we ought to change the way we treat robots, however it highlights a bias hidden in the process.

We arrive at a central finding in our investigation, namely, whether or not the distinction between human others and robots is an artifact. The very phenomenon of our becoming aware of a robot simulating something, and in particular, when robots appear to try to be something that they "clearly" are not, facilitates a rare glimpse behind the epistemological scenes. Our concerns about robotic simulation phenomenologically reveal the general conditions of cognition, of how we construct the world around us, and ultimately, ourselves. Our way of interpreting the world is rather limited; we are constantly inferring and deducing from a conditioned standpoint, and using this to understand and categorize. These limitations amount to conceptual distortions/approximations/interpretations (for better and for worse) of the original experiential "content" received through our limited input channels, and thus, the distinction between a human being and a humanoid robot, for example, is literally an artificial construct.

Our concerns about robotic *simulation*, which involve elements of the inverse of discrimination, reveal, or so I have suggested here, the much more general epistemological *discrimination* involved in any conceptualization. Once this is realized, our concerns about robotic *simulation* become a case of "lived epistemology," where epistemological conditions suddenly show up in concrete contexts of everyday life. Given these perspectives, can HRI provide us with the opportunity to disclose insights about the human part involved in the encounter? This question will be addressed in the next sections.

Before presenting a different conceptual design approach that aims for a praxis with robots in the mode of BA1 (Section 6), it is important to explore the consequences that are latently bound to the modes in which we further engage with the world. In the next section, I relate the above-sketched theory of *Bodymind Awareness* to Heidegger's thinking on how technology engages us in the world. Here, I draw on the *Question Concerning Technology* (2004), as it is rather compatible with the thought presented above, and hints at the danger and potential of modern technology, which I relate to HRI.

Dangers and Potential

In the *Question Concerning Technology* (2004), Heidegger famously discusses the profound changes that technology has imposed on modern civilizations. These are not minor, superficial, cultural changes. They affect the very nature of human beings, which Heidegger describes in terms of interactions. Modern technology involves humans in a new and problematic mode of interaction, which he characterizes as follows.

> Man stands so decisively in attendance on the challenging-forth of Enframing that he does not apprehend Enframing as a claim, that he fails to see himself as

the one spoken to, and hence also fails in every way to hear in what respect he ex-sists, from out of his essence, in the realm of an exhortation or address, and thus can never encounter only himself. (Heidegger 2004, p. 46)

What Heidegger is referring to, here, in his own peculiar language, may perhaps be best explained by simultaneously referring back to the case of K, in Ambo's documentary, introduced previously. Modern technology offers us the ability to address phenomena as *standing-reserve* (*Bestand*), which makes them into always-available resources or commodities (for details see ibid., p. 41). K's high-tech robot companion, PARO, is always available to K (of course PARO has to be maintained—the battery has to be charged, etc.), and the unity of all that constitutes PARO as an entity has become a *standing-reserve* to her. Our willingness to consider some robot interaction partners as *standing-reserves* may also be seen in a study of Sony's entertainment robot dog, AIBO, conducted by Kahn, Friedman, and Hagman (2002) who analyzed online discussion forums about AIBO, and concluded:

> Results showed that AIBO psychologically engaged this group of participants, particularly by drawing forth conceptions of essences (79%), agency (60%), and social standing (59%). However, participants seldom attributed moral standing to AIBO (e.g., that AIBO deserves respect, has rights, or can be held morally accountable for action). (ibid., p. 632)

Essence, *agency*, and *social standing* were all ascribed to AIBO, but this did not assign any *moral standing* to the robot. I take this to indicate that we do not consider ourselves to have any obligations towards AIBO, and as a consequence, that we may use it in any way that suits our purpose. The ascription of *moral standing* would probably also make the treatment of AIBO as a *standing-reserve* unnatural.

> We now name that challenging claim which gathers man thither to order the self-revealing as standing-reserve: 'Ge-stell' [Enframing]. (Heidegger 2004, p. 42)

Hence, the PARO Enframing (*Ge-stell*) is what tempts K to perceive PARO as a *standing-reserve*, ready for usage as such, whenever it pleases her. Hence, The first Heidegger quote in this section warns against the latent danger accompanying modern technology, namely, the unawareness of this process as it happens. K is, so to speak, faced with the risks of:

1. Not seeing herself as being addressed by the Enframing, which may absorb her fully in the HRI.
2. Never meeting herself outside the context set up by the Enframing.

Summarized, the utmost danger is becoming fully absorbed into the *Enframing* of a robot, for example, and its consequent reduction to a *standing-reserve*. It is this kind of estrangement in the interaction against which Heidegger warns us.

In the *Question Concerning Technology*, Heidegger implicitly presupposes that technological advance is unstoppable. However, besides cautioning us against its dangers, he also sees great possibilities:

> As this destining, the coming to presence of technology gives man entry into That which, of himself, he can neither invent nor in any way make. For there is no such thing as a man who, solely of himself, is only man. ... it is precisely in this extreme danger that the innermost indestructible belongingness of man within granting may come to light, provided that we, for our part, begin to pay heed to the coming to presence of technology. (ibid., p. 49)

Briefly put, technology contains the potential for the rediscovery of interaction that constitutes our essence, our humanity. In the course of interaction with a robot, we have, on the one hand, the danger of reducing it to the notion of *standing-reserves*, however, on the other hand, it also facilitates the possibility of becoming aware of our "highest dignity," which is:

> This dignity lies in keeping watch over the unconcealment—and with it, from the first, the concealment—of all coming to presence on this earth. (ibid., p. 49)

Therefore, the opportunity that technology offers us through *Enframing* lies in the realization we are granted by being connected to the *Unconcealed* (*Unverborgenheit*), as well as the *Concealed* (*Verborgenheit*), and in the possibility of becoming involved in a different mode of *Revealing* (*Entbergen*) the *Unconcealed*, than as *standing-reserves*.

Bodymind Awareness in the Context of Technology

Heidegger's terminology is quite idiosyncratic, and his considerations are exceedingly abstract, so in the first instance, his interpretation of the essence of technology is likely to raise questions of understanding. Attempting to relate the three different modes of *Bodymind Awareness* to his theory, it seems that BA1 grants us access to the *Unconcealed*.

> We need only apprehend in an unbiased way that which has already claimed man and has done so, so decisively that he can only be man at any given time as the one so claimed. Wherever man opens his eyes and ears, unlocks his heart, and gives himself over to meditating and striving, shaping and working, entreating and thanking, he finds himself everywhere already brought into the unconcealed. (ibid., p. 42)

Recall that in BA1 we do not force any *noetic vector*. Furthermore, it is important to note that this mode does not necessarily conflict with being engaged in activities. Hence, BA1 seems to realize the openness described in the quote above, from the *Question Concerning Technology*. This mode of being with the phenomena, a *letting be* (*sein lassen*), will be further elaborated in the next section, as a mode of being with robots that does not reduce the robot to a *standing-reserve*.

In contrast to BA1, BA2 and BA3, strap phenomena to our categorical apparatus, and at these levels we find the danger of being consumed by the *essencing* (*wesen*) of technology. Hence, it is necessary to take the previously-stated warnings seriously, as here, by the reduction of the robot to a mere *standing-reserve*, we risk exposing ourselves completely to the mandate of technology, as in the case of K, above. Nevertheless, it is also here that we can try to make sense of Heidegger's general claims about the "saving potential" of technology at a certain stage of its development. As also argued above, the perception of robots may be a special type of experience, as it may thoroughly connect us to our epistemological boundaries. Categorical inconsistency may force us to realize the tension between BA1 and BA2/BA3, namely, our capacity to be intentionally and experientially open to reality as such (BA1), which both presents itself and conceals itself when we try to conceptualize experiential content (BA2, BA3). By doing so, we become aware of the *essencing* of technology, by being prompted to reflect existentially on the encounter with the given robot. Thereby will we face the problem that our classifications may be unreliable, and consequently, our self-understanding as free agents is suddenly on shaky ground—a construct. This conclusion actually resonates with the general findings of cognitive science, namely, that the self is a illusion (cf. Hood 2013, Frith 2007). In the type of HRI outlined above, in some sense we are already saved from the Heideggerian danger of technology, as this lies in not being aware of its *essence* (cf. Olsen 2013, p. 135).

In the next section, I offer a constructive approach to robots, a mode of interacting with them that goes beyond being aware of the *essence* of technology only, by bypassing the danger from the very beginning.

HRI and Human Self-Realization

In this section I suggest that we combine Heidegger's general insight into the constitutive power of "right" interactions with insights by Buddhist-inspired epistemology as presented in the previous sections.

I have just argued that the positive aspect of technology lies in the fact that at some point in its development, a kind of artifact is produced that empowers us to realize our epistemological situation. Hence, in the interaction with robots, we have the opportunity to recognizing the *essence* of technology, and authenticating ourselves by adopting another, non-technological, "mode of revealing." It is the latter that will be of concern in here. This other "mode of revealing" far exceeds the adoption of some ethical principle that would apply moral obligations to robots for this or that meta-ethical reason. Meeting PARO and other service robots carries a potential that goes beyond the decision to refrain from reducing robots to a commodity for consumption, or to cite Heidegger, a *standing-reserve*. Instead, it consists of adopting the standpoint of enablers of experience in the sense of BA1, as the totality of an interaction.

Reflections on the "right" way of *being-with-robots* could begin with a look at the function of praxis. Through praxis we meet the world, in some sense improve and evolve, and we can learn to remain in BA1 in the moment of interaction. In this regard, Japanese fine arts such as *kabuki* theatre, the martial arts, and the tea ceremony, are considered as conditioning higher forms of mind-body integration (Shaner 1985, pp. 99-100). The unifying approach to praxis creates some sort of integration, rather than discrimination among all participants, and could be interpreted as enabling a form of self-realization by fostering prolonged interactions at the level of BA1. The rationale behind terming it "self-realization" is that it is in BA1 that we can come the closest to not only meeting robots, but also ourselves, in our suchness. As a consequence, taking the quotation in the introduction by Sparrow and Sparrow seriously, we might even say that it is also our duty to engage in this kind of activity.

Since in principle, "integrated forms of praxis" seem to be possible with any object, is there any reason why robots should not be able to take the role of the bow, flower, or tea? Service robots, like vacuum cleaners, are developed to satisfy a concrete demand. However, the praxis they are thought to support is most often not developed to go beyond being a traditional means to an end. For instance, we push a button to start the cleaning robot, in order to accumulate time that we can spend elsewhere. We try to satisfy our needs, for example, by letting a robot do the work, and in the case of more advanced robots, by interacting in a more or less sophisticated manner with them. Nevertheless, the type of robotic use will likely differ from the conditioning practice described above. A step towards a novel (additional) form of HRI seems to be to design service robots in such a way that they afford human involvement as "integrated forms of praxis" that converge our intentionality towards B1, and hence beyond our limited conceptual boundaries.

Admittedly, these are still vague and incipient considerations, but I trust that they at least indicate the general direction of how to more concretely develop these thoughts. By focusing on integrated forms of praxis, we would minimize the risk of a deliberate exploitation of robots, and the corresponding devaluation of ourselves within the context of the framework provided. The interaction between a human and a robot could suddenly achieve a new dimension, akin to the artist's interactions with his surroundings during the creation of a work of art. In this regard, inventions such as the *Partner Ballroom Dance Robot*, developed at Tohoku University, may be a promising approach to robots, affording a praxis similar to that outlined above.

Concluding Remarks

In the previous sections, inspired by Buddhist thought, I have argued that our way of conceptualizing dehumanizes our environment, and as a consequence, turns robots into artifacts. Moreover, I have indicated that interactions with and through robots could have significant epistemic potential for us, namely, enabling the possibility of getting to know oneself better, as BA1 may give us the most direct access to reality, of which we are part. As this is a practical process, it may be considered a form of self-realization.

Of course, if we design robots badly, or with malicious intent, it may be hard to enrich our lives by interacting with them. The foregoing does not overrule ethical considerations, nor suspend them, as such. However, my elaborations should have plausibly presented the possibility of there also being a different mode of interacting with robots that will not necessarily reduce us to the mere mandate of the robot, and which has, to my knowledge, so far been overlooked by the literature. Here, we find more than just the mitigation of the *danger* pointed out by Heidegger, as this sort of praxis may even help us to realize a neglected potential of HRI—namely, the opportunity cultivate ourselves, by opening up to the phenomenon as such. An interesting question for further research would be whether robots are, or could be, special enablers in this regard. We have probably never before seen technologies that make us reflect to such an extent on who we are, and simultaneously trigger an urge in us to differentiate ourselves categorically from the essence of its instances.

One may object that existing robots do not invite humans to engage with them in the type of interaction described, and that this type of robot will probably never be more than a minor phenomenon, owing to various economical agendas or technical limitations. This may indeed be the case. Nevertheless, there seems to be a possibility of self-realization through robots. Drawing attention to this potential is different from suggesting that every encounter with a robot should and would stimulate some sort of epiphany.

References

Abe, M. 1985. *Zen and Western Thought*. Edited by William R. LaFleur. Honolulu: University of Hawai'i Press.

Abe, M. 2003. *Zen and the Modern World: A Third Sequel to Zen and Western Thought*, edited by Steven Heine: University of Hawai'i Press.

Akira, H. 1990. *A History of Indian Buddhism*. Translated by Paul Groner, edited by Paul Groner. Vol. 36, *Asian Studies at Hawaii*: University of Hawaii Press. Original edition, Indo Bukkyou shi.

Ambo, P. 2007. *Mechanical Love*. Denmark: Dox On Wheels.

Bein, S. 2011. *Purifying Zen: Watsuji Tetsuro's Shamon Dogen*. Honolulu: University of Hawai'i Press.

Calverley, D.J. 2006. "Android Science and Animal Rights, Does an Analogy Exist?" *Connection Science* 18(4): 403-17. doi: http://dx.doi.org/10.1080/09540090600879711.

Cousins, L.S. 1998. Buddha. In *Routledge Encyclopedia of Philosophy*, edited by E. Craig. London: Routledge.

DaneAge. 2009. "Ja Tak Til Robotter I Ældreplejen." *Ældre Sagen NYT*, June.

Det Etiske Råd. 2010a. "Cyborg Teknologi: Udtalelse Fra Det Etiske Råd." Accessed January 18, 2010. www.homoartefakt.dkDet Etiske Råd. 2010b. Recommendations Concerning Social Robots. The Danish Council of Ethics.

Det Etiske Råd. 2010b. "Recommendations Concerning Social Robots." The Danish Council of Ethics Accessed October 10, 2014. http://etiskraad.dk/Temauniverser/Homo-Artefakt/Anbefalinger/Udtalelse%20om%20sociale%20robotter.aspx?sc_lang=en.

Det Etiske Råd. 2010c. "Sociale Robotter: Udtalelse Fra Det Etiske Råd." Accessed January 18, 2010. www.homoartefakt.dk.

Frith, C. 2007. *Making up the Mind: How the Brain Creates Our Mental World*: Blackwell Publishing.

Greenfieldboyce, N. 2008. "Robotic Baby Seal Coming to U.S. Shores." NPR Accessed 2015. http://www.npr.org/templates/story/story.php?storyId=91875735.

Gunkel, D.J. 2012. *The Machine Question: Critical Perspectives on AI, Robots, and Ethics*. Cambridge, MA: The MIT Press.

Heidegger, M. 2004. "Question Concerning Technology." In *Readings in the Philosophy of Technology*, edited by David M. Kaplan. Oxford: Rowman & Littlefield Publishers, Inc.

Heisenberg, W. 1927. "Über den anschaulichen Inhalt der quantentheoretischen Kinematik und Mechanik." *Zeitschrift für Physik A Hadrons and Nuclei* 43(3-4): 172-98. doi: http://dx.doi.org/10.1007/BF01397280.

Hood, B. 2013. *The Self Illusion: How the Social Brain Creates Identity*. New York: Oxford University Press.

Ishiguro, H. 2006. "Android Science: Conscious and Subconscious Recognition." *Connection Science* 18(4): 319-32. doi: http://dx.doi.org/10.1080/09540090600873953.

Kahn, P.H., B. Friedman, and J. Hagman. 2002. ""I Care About Him as a Pal": Conceptions of Robotic Pets in Online Aibo Discussion Forums." *Extended Abstracts of CHI 2002 Conference on Human Factors in Computing Systems*:632-3.

Kasulis, T.P. 1985. *Zen Action, Zen Person*. Honolulu: University Press of Hawaii.

Kitano, N. 2005. "Roboethics: A Comparative Analysis of Social Acceptance of Robots between the West and Japan." *The Waseda Journal of Social Sciences* 6: 93-105.

Lovgren, S. 2007. "Robot Code of Ethics to Prevent Android Abuse, Protect Humans." *National Geographic News*, March 16.

Marti, P., A. Pollini, A. Rullo, and T. Shibata. 2005. "Engaging with Artificial Pets." *papers://74134E92-DC57-45CA-BDE4-6B8D82D9C6E7/Paper/p814*, EACE

'05: Proceedings of the 2005 annual conference on European association of cognitive ergonomics, Sep 1. 99-106.

Mejor, M. 1998. Suffering, Buddhist Views of Origination Of In *Routledge Encyclopedia of Philosophy*, edited by E. Craig. London: Routledge.

Mori, M., K.F. MacDorman, and N. Kageki. 2012. "The Uncanny Valley." *IEEE Robotics & Automation Magazine* 19(2): 98-100. Original edition, 1970. doi: http://dx.doi.org/10.1109/MRA.2012.2192811.

Nadeau, J.E. 2006. "Only Androids Can Be Ethical." In *Thinking About Android Epistemology*, edited by Kenneth M. Ford, Clark Glymour and Patrick J. Hayes, 241-8. Menlo Park: AAAI Press.

Nishida, K. 1990. *An Inquiry into the Good*. Translated by Masao Abe and Christopher Ives: Yale University Press. Original edition, 1911.

Nørskov, M. 2011. "Prolegomena to Social Robotics: Philosophical Inquiries into Perspectives on Human–Robot Interaction." PhD dissertation, Department of Philosophy, Aarhus University.

Nørskov, M. 2014a. "Human–Robot Interaction and Human Self-Realization: Reflections on the Epistemology of Discrimination." In *Sociable Robots and the Future of Social Relations: Proceedings of Robo-Philosophy 2014*, edited by Johanna Seibt, Raul Hakli and Marco Nørskov, 319-327. Amsterdam: IOS Press Ebooks. doi: http://dx.doi.org/10.3233/978-1-61499-480-0-319.

Nørskov, M. 2014b. "Revisiting Ihde's Fourfold "Technological Relationships": Application and Modification." *Philosophy & Technology*:1-19. doi: http://dx.doi.org/10.1007/s13347-014-0149-8.

Olsen, S.G. 2013. "Hvad Er Det Væsentlige Ved Teknikken? En Ny Læsning Af Heideggers Foredrag." In *Nye Spørgsmål Om Teknikken*, edited by Kasper Schiølin and Søren Riis, 123-37. Aarhus: Aarhus University Press.

Politiken. 2008. "Danskerne Er Bange for Robotter." Politikken.dk Accessed 6 January. http://politiken.dk/videnskab/ECE567229/danskerne-er-bange-for-robotter/.

Regan, T. 2005. *Empty Cages: Facing the Challenge of Animal Rights* Rowman & Littlefield Publishers Inc.

Rodogno, R. 2014. "Social Robots and Sentimentality." In *Sociable Robots and the Future of Social Relations: Proceedings of Robo-Philosophy 2014*, edited by Johanna Seibt, Raul Hakli and Marco Nørskov, 241-4. Amsterdam: IOS Press Ebooks. doi: http://dx.doi.org/10.3233/978-1-61499-480-0-241.

Shaner, D.E. 1985. *The Bodymind Experience in Japanese Buddhism*, edited by Kenneth Inada, *Suny Series in Buddhist Studies*. Albany: State University of New York Press.

Sharkey, A. and N. Wood. 2014. "The Paro Seal Robot: Demeaning or Enabling?" AISB 2014 - 50th Annual Convention of the AISB, UK.

Smart, N. 1997. The Buddha. In *Companion Encyclopedia of Asian Philosophy*, edited by Brian Carr and Indira Mahalingam. London: Routledge.

Sparrow, R. 2002. "The March of the Robot Dogs." *Ethics and Information Technology* 4(4): 305-18. doi: http://dx.doi.org/10.1023/A:1021386708994.

Sparrow, R. and L. Sparrow. 2006. "In the Hands of Machines? The Future of Aged Care." *Minds and Machines* 16(2): 141-61. doi: http://dx.doi.org/10.1007/s11023-006-9030-6.

Takayama, L., W. Ju, and C. Nass. 2008. "Beyond Dirty, Dangerous and Dull: What Everyday People Think Robots Should Do." *HRI 2008 Proceedings of the Third ACM/IEEE International Conference on Human–Robot Interaction*, 3rd ACM/IEEE International Conference on Human–Robot Interaction, Amsterdam, The Netherlands, 12-15 March 2008. 25-32.

Turkle, S. 2012. *Alone Together: Why We Expect More from Technology and Less from Each Other*. New York: Basic Books. Original edition, 2011.

Veruggio, G. 2006. "The EURON Roboethics Roadmap." 2006 6th IEEE-RAS International Conference on Humanoid Robots, Genoa, Italy, 4-6 December 2006 612-17 doi: http://dx.doi.org/10.1109/ICHR.2006.321337

Chapter 7

The Uncanny Valley:
A Working Hypothesis

Adriano Angelucci, Pierluigi Graziani and
Maria Grazia Rossi[1]

In the following chapter we put forward a working hypothesis concerning the so-called *uncanny valley* phenomenon. This technical term refers to the familiar feeling of revulsion that we typically experience while visually inspecting human-looking subjects in which something does not seem to fall in the right place. Insofar as this phenomenon prevents us from engaging in empathic interactions with androids, humanoid robots or digital characters, it has been studied almost exclusively within the fields of robotics and computer graphics, where it poses a mainly *technological* challenge. After reviewing different ways of overcoming this technological challenge, we claim that the same phenomenon also lies at the heart of a further, *philosophical* challenge, originating in our reluctance to recognize characters falling in the uncanny valley as humans. Our working hypothesis is that the notion of *dehumanization* may constitute an illuminating conceptual bridge connecting the emotion of *disgust*, broadly construed, to the uncanny phenomenon.

Familiarly "Strange" Feelings

In 1906, the German psychiatrist Ernst Jentsch first made use of the German term *unheimlich*, "uncanny", in order to characterize a strange feeling of "unnaturalness" often experienced by humans while observing situations or pictures of human-looking subjects in which something does not seem to fall

1   It is only fair to acknowledge that, while the sections *Familiarly "Strange" Feelings* to, *The Long Reach of the Uncanny*, and *A Technological Challenge* of the present paper are the result of the joint effort of its three authors, *A Philosophical Challenge and a Working Hypothesis* is due in particular to Maria Grazia Rossi. An earlier version of this work appeared as Angelucci, A., M. Bastioni, P. Graziani, and M.G. Rossi. 2014. "A Philosophical Look at the Uncanny Valley." In *Sociable Robots and the Future of Social Relations: Proceedings of Robo-Philosophy 2014*, edited by Johanna Seibt, Raul Hakli and Marco Nørskov, 165-71. Amsterdam: IOS Press Ebooks. doi: http://dx.doi.org/10.3233/978-1-61499-480-0-165. Republished with permission of IOS Press.

in the right place. In 1919, the father of psychoanalysis, Sigmund Freud (Freud 1953), set out to inquire into the causes of the same phenomenon, but arguably failed to reach a complete understanding of the matter, due to a self-confessed lack of familiarity with the very phenomenon under investigation. Both authors characterized the peculiar feeling as a strange sensation, not necessarily of repulsion, typically elicited by episodes of *déjà vu* or by situations in which reality seems to fade into a dream.

Half a century later, the same phenomenon began to be systematically investigated. In 1970, the Japanese roboticist Masahiro Mori made use of the locution "Bukimi no Tani Genshō" (literally "the phenomenon of the uncanny valley") in order to refer to the eerie sensation (i.e. *bukimi*), mostly of revulsion, often generated in us by robots carefully designed to look human. The now technical expression *uncanny valley* refers to a remarkable feature of the graph originally used by Mori in his pioneering study. Relying on anecdotal data, Mori plotted *familiarity* (i.e. *shinwakan*) as a function of *human likeness*. As Mori's picture shows,[2] familiarity and human likeness seem to stand in an almost linear relation up to a certain point, beyond which even a slight increase in human likeness results in an abrupt and drastic decrease in familiarity, followed by an equally drastic increase of the same variable as human likeness keeps increasing and approaches perfection (drawing the *valley*).

Mori's graph provides a very important piece information. As the graph shows, since *still* images can never reach a high level of familiarity, the corresponding uncanny feeling they may elicit is normally experienced as less intense than the one elicited by *moving* images. A moving zombie, for instance, generates a deeper valley than a still corpse.

Whereas Mori's graph has the merit of facilitating an intuitive understanding of the uncanny, the phenomenon it represents is in fact far more complex than it may suggest on a first look. The major complication derives from the fact that the curve plotted in Mori's graph seems to be highly *adaptive*. Indeed, as subsequent research has shown, the more human likeness increases, the more efficient our brains become at detecting small details of the image, capable of eliciting uncanny feelings (Gouskos 2006, Tinwell, Grimshaw, and Williams 2011). As a consequence, every time one of these details is uncovered and taken care of, other, more fine-grained ones are spotted, and the graph needs to be corrected accordingly (see the last Section of this Chapter).

As we shall see, the uncanny phenomenon poses two different challenges, a technological as well as a philosophical one. We will start by considering the former. This will allow us to introduce some elements needed to adequately analyze the latter.

---

2   An English translation of Mori's article from 1970 (in Japanese), which also contains his graph, can be found in The Uncanny Valley by Mori, MacDorman, and Kageki (2012).

The Long Reach of the Uncanny

While it has only recently began to be systematically studied, the uncanny valley phenomenon (henceforth simply *uncanny*) seems to go way back in human history. In fact, some authors connect it to a primitive fear of death, presumably already familiar to our hunter-gatherer ancestors (see MacDorman's papers in bibliography). Yet the advent of the so-called information age seems to have powerfully contributed to enhance its effects. As a consequence, different areas of research have started to develop ways to deal with it.

Less than a century ago expressions such as "android" or "digital character" had no meaning at all for the man in the street. Yet things change rapidly. At the time of writing, billions of human–computer interactions presupposing these notions are taking place on our planet. These range from more obvious ones, such as using smart phones or playing video games, to less apparent and more subtle ones, such as watching movies realized in computer graphics or being bombarded by commercials cunningly created for various kinds of digital media. Unsurprisingly, the uncanny has correspondingly started to represent a potential source of troubles for computer scientists developing all kinds of user-friendly interfaces, in that it can disrupt their attempts at reaching the level of empathy required for successful man-machine interactions. Indeed, insofar as these interactions rely crucially on computer-generated imagery, they can be easily haunted by unforeseen and unwelcome uncanny effects. In the process of creating new PC help desk interfaces, for instance, programmers have long learned to steer clear of synthetic humans, in order to avoid the aversion typically generated by syncopated speech patterns *à la* Max Hedroom.

The uncanny is also systematically exploited by the film industry. Indeed, by cunningly altering tiny and apparently insignificant physiognomical traits, horror filmmakers can easily generate in their audiences a disquieting sensation of fear and repulsion. On the other hand, the very existence of an uncanny can also have far-reaching economic consequences, and can even contribute, at times, to the failure of large scale business ventures. In this regard, an eye opening example of the tangible risks one can expose oneself to by underestimating its importance comes from 3-D Digital Cinema. It is a known fact, that *The Polar Express*, the first all-digital capture film, written, produced, and directed by Robert Zemeckis, went dangerously close to becoming a box office bomb on its first release in November 2004. The film featured human characters animated by means of performance capture techniques. Differently from what their creators expected or hoped, these characters were widely criticized for being, "creepy", "spooky", "fake-looking", and "mannequin-like" ... in other words, for falling squarely in the uncanny (see for example *The Economist* 2010, Tinwell's papers in the bibliography and MacDorman and Ishiguro 2006).

The film industry, however, is far from being the only field in which avoiding the uncanny has proved necessary. Arguably, the era of virtual environments in which geographically distant people can meet and interact in a realistic way is only in its infancy. As it often happens, the technological means used to breathe life into these carefully crafted synthetic worlds owe much to the video game industry. Indeed, role-playing games in which client-men interact and evolve in server-worlds have now long exited. It does not take much effort to realize that the existence of virtual spaces has already begun to reshape the very way in which new generations conceive of social relations, including love affairs. Yet the full potential of virtual environments remains largely unexplored. Work, social activities, education, healthcare, and psychological counseling taking place in wholly virtual spaces are all very live possibilities, and in each of these computer-mediated interactions the looming of the uncanny could be seriously disruptive.

For obvious reasons, robotics is perhaps the field in which the uncanny has received the most attention. Indeed, scenarios in which androids assist men in the most disparate activities still sound futuristic, but may not be too far away. An *android* is very different from a humanoid robot, for which similarity to humans is not considered fundamental, and it has recently been characterized in functional terms as "an artificial system designed with the ultimate goal of being indistinguishable from humans in its external appearance and behavior" (MacDorman and Ishiguro 2006, p. 289). To the extent that the ultimate goal of an android creator is that of reaching a perfect human resemblance, the uncanny stands clearly on its way to success. The fact that the most famous Japanese robotic realizations tend to avoid realism and to make use of very stylized, hi-tech shapes, for instance, can hardly be seen as a coincidence. Only recently, thanks to the work of Hiroshi Ishiguro and his team at the *Intelligent Robotics Laboratory*, there have been partly successful attempts at creating human-looking robots. Unfortunately, as of today, the creation of an android truly "indistinguishable from humans in its external appearance and behavior" is still not possible in practice. True, the use of organic-looking materials to realize skin and hair (already familiar to the *animatronics* industry of the Eighties) could give the android a very realistic look. But this is only insofar as the android does not move. Once in motion, the lack of the about 600 muscles which compose a human body, of their fluidity and harmony, is immediately detected by a human observer and the uncanny effect appears.

On a more theoretical level, one of the more interesting uses of androids has recently been championed by Karl MacDorman and Hiroshi Ishiguro (2006). Their research, in particular, has emphasized the close connection between social sciences and android development. Social, cognitive, and neuroscientific models, in their view, can all be explored by using androids in the place of human actors. It seems only fair to acknowledge that their prospects, while highly speculative, are undeniably fascinating. While we traditionally needed to study social and cognitive sciences in order to build robots that behave like humans, we would now, according to these authors, be in a position to reverse this trend, and make use of robots in order to better understand psychological and social dynamics. What makes this possible, they believe, is the fact that androids typically elicit a person's model of a human other. In particular, precisely *because* of the uncanny, androids may turn out to provide the best means of finding out what kinds of behaviors we perceive as human. Indeed, according to MacDorman and Ishiguro:

> An experimental apparatus that is indistinguishable from a human being, at least superficially, has the potential to contribute greatly to an understanding of face-to-face interaction in the social, cognitive, and neurosciences. It would be able to elicit the sorts of responses, including non-verbal and subconscious responses, that people typically direct toward each other. Such a device could be a perfect actor in controlled experiments, permitting scientists to vary precisely the parameters under study. The device would also have the advantage of physical presence, which simulated characters lack. (ibid., p. 298)

It should not be necessary to add, on the other hand, that, pending a full understanding of the uncanny, a similar research program can hardly get off the ground. In fact, MacDorman and Ishiguro intend to base their new experimental paradigm on a full-blown *android science*, which they define as the science that studies "the significance of human likeness in human-machine relationships" (ibid., p. 302).

The very complexity of the phenomenon, in our view, suggests that robotics may not be today the best way of exploring the uncanny. A better instrument to investigate it, as well as to experiment with the many different features the make us look human, we believe, may be found in the thriving field of *computer graphics*.

A Technological Challenge

One reason we have for favoring this instrument is that computer graphics allow us to easily overcome the major financial problem represented by the enormous costs related to the construction of a physical robot. Whereas the famous robot clones of Hiroshi Ishiguro, for instance, which makes use of only 50 actuators (consisting of miniature hydraulic pistons), costs about one million dollars, one can simulate hundreds of muscles within the body of a computer-generated character by means of a much more affordable software. As a consequence, the already mentioned application of computer graphics in the film and video game industry is likely to make funding easier to secure.

Financial considerations aside, there is a host of other practical reasons for preferring digitally created characters over physical robots. Indeed, besides their being free, for obvious reasons, from hardware-related problems, they can be shared by researchers and research teams working at great distances. As any other software, they can be improved much more easily than robots, which need to be rebuilt each time. Last but not least, they are much simpler to interface with motion capture systems, in that they do not have to take into account problems such as gravity or balance.

This being said, computer graphics still has its own obstacles to overcome. In particular, two main approaches have been adopted so far in the development of digital characters. Whereas some try to simulate from scratch human anatomical features such as bones, muscles, tendons, skin, and the like by means of mathematical models of increasing complexity, others find more promising to digitally capture the expressions and motions of real human subjects, store them in enormous databases, and reproduce them at will (Bastioni and Graziani 2011). Theoretically, the former would be the most interesting way of reaching complete realism. If it were possible to devise mathematical models by means of which one could perfectly simulate the external appearance of a human being, the uncanny would thereby cease to represent a problem. Unfortunately, the problems of this approach are at present insurmountable. The next two subsections will give the reader an idea of the kinds of difficulties besetting each approach.

The Elusiveness of a Smile: Simulating Reality

Let us focus on one of the most common cross-cultural facial expressions: smiles. The following considerations are intended to give an idea of the many difficulties involved in generating a smiling digital character.

Facial muscles are over 30, about a dozen of which are involved in smiling. The contraction of each of these muscles during a smile varies greatly in speed, intensity, and direction, it hinders or facilitates the contraction of nearby muscles, and propagates to facial cartilages. This complex pattern of forces is then transmitted to the skin, the movements of which critically depend on the amounts of fat present between skin and muscle. As a consequence of this complexity, skin behavior can easily become unmanageable. Compressing and stretching the skin alters the blood flow and gives rise to different skin complexions, thereby making blood vessels more or less visible. Depending on elasticity, collagen percentages, and other concurrent factors, including highly subjective ones such a habitual facial expressions, wrinkles, and (possibly) scars may also become more or less visible. Stretched skin, on the other hand, tends to make facial bones more visible. All of these tiny changes, moreover, have a crucial impact on the reflectance properties of the skin surface. Indeed, light does not just bounce off the skin, but penetrates into it. When it eventually reaches the muscles, more dense and red, it is scattered back in the most unpredictable ways, thereby lending the skin its typical pinkish hue (in the technical jargon, this behavior is usually referred to as *surface scattering*). These reflectance properties may also be altered by numerous other more or less external factors, such as facial hair, makeup, perspiration or rain.

As we all know, a human face is an astonishing kaleidoscope of expressions, each one of which fades quickly yet continuously into another, to the effect that the boundary between a smile an a grimace may become surprisingly thin. Yet our brain is notoriously the ultimate pattern recognition machine. From very early stages of our development, it somehow manages to keep this extraordinary abundance of information under control. As a consequence of this hard-wired prowess, the neglect of even one small detail in our computer simulation can determine unwanted uncanny effects. This is due to the fact that our recognition of a human face as such largely depends on a series of subconscious expectations. Different areas of the face are expected to move in certain ways and to have specific colors, which in turn are expected to change in response to various internal and external conditions. The behavior of a drop of sweat on a face has to exactly match our most fine-grained expectations, if it does not, the fragile illusion is immediately shattered.

For all we know at present, the mathematical models necessary to simulate all this are very likely to be unimaginably complex (Bastioni and Graziani 2011). Current research focuses mostly on muscles behavior, which is relatively easier to simulate. The leading trend is that of trying to simulate particular facial expressions by combining basic muscle contractions. Some researchers are also working on skin elasticity and soft tissues behavior. At times attempts are made to join efforts in the creation of a lifelike digital human face, but the results are still far from sufficient to avoid the emergence of uncanny effects. In fact, due to the apparent lack of naturalness of the characters thereby obtained, the industry still prefers to rely on motion capture techniques, in which real subjects faces are covered with sensors, and their movements carefully recorded. This takes us to the aforementioned second approach that has been adopted so far in the creation of digital characters, which will be considered in the next subsection.

The Elusiveness of a Smile: Capturing Reality

According to many, given the rapid increase in computers storage capacity, the best way to avoid the uncanny would consist not so much in trying to *simulate* reality, but rather in devising reliable methods to *capture* it. Going back to our smile, we could put aside mathematical models altogether, resort to a smiling actor in the flesh, and start to accurately record and store each of her facial movements, down to the most involuntary microexpressions. This would allow us to set up a database consisting of gigabytes of three-dimensional coordinates, each of which would correspond to the position of a specific point of her face. If we further lit the actor's face from every possible lighting direction, we could be able to record and store the surface reflectance properties peculiar to her skin and set up a further database, this time consisting of gigabytes of images. The information stored in these databases would finally allow us to create a digital clone of the physical actor and to animate the digital character in order to make it act in a lifelike fashion, as in a conversation for instance.

In 2008, thanks to the groundbreaking work of computer scientist Paul Debevec, *Image Metrics*, a company specialized in facial animation techniques for video games and films, together with the *USC Institute for Creative Technologies* created *Digital Emily*, an astonishingly perfect digital clone of the actress Emily O'Brien, and realized an impressive virtual interview in which the digital character itself explained the technology used for its animation (Alexander et al. 2009, Alexander et al. 2013). In commenting on the extraordinary achievements of the so-called Emily Project, the writer, digital artist and software designer Peter Plantec recently wrote: "I officially pronounce that Image Metrics has finally built a bridge across the Uncanny Valley and brought us to the other side" (Plantec 2008). The technology inaugurated with the realization of Digital Emily, being able to keep the uncanny relatively under control, has already found applications in the film industry.[3] The next challenge consisted in applying the new film technology to the video game world. This transition is not nearly as easy as it may seem. Indeed, digital film making has a lot of time to render its characters. A few minutes digital scene can normally count on weeks of work and expensive hardware in order to be realized. The digital character of a video game, on the other hand, has to run on popular video game consoles, and, most importantly, has to respond in real time to the player's inputs. Despite these difficulties, in 2012, a collaboration between *USC Institute for Creative Technologies* and *Activision*, also led by Paul Debevec created *Digital Ira*. Whereas Digital Emily is a pre-computed simulation rendered offline, Digital Ira is a photoreal digital character that runs in real-time. Just as its predecessor, Digital Ira seems to fare well with respect to the uncanny, and to keep it relatively at bay.

This being said, attempts at fully capturing reality have only just began and still have a long way to go. Nonetheless, while technical considerations suggest that the bridge across the uncanny valley envisioned by Plantec may still not be as solid as one would wish, the technological challenge posed by the uncanny seems on its way to being overcome. Yet a deeper theoretical challenge still needs to be faced.

A Philosophical Challenge and a Working Hypothesis

As the above considerations should have made sufficiently clear, at least within the fields of robotics and computer graphics the admittedly tough challenge posed by the uncanny is (perhaps understandably) perceived as mainly technological in nature. This challenge, as we saw, has to do with the problem of devising ingenious technologies capable of reaching increasing levels of pictorial realism. Accordingly, let us refer to the problem that these technologies are intended to solve as *the problem of realism*. From this last point of view, the uncanny is caused by a lack of microdetails. As we saw, this lack is responsible for our perceiving humanoid bodies and their movements as unnatural.

We believe that the problem of realism, although certainly important, may not be all there is to the uncanny. More specifically, in this last section we would like to shift the attention of the reader from that which *causes* the uncanny

---

3   See, for example, the ICT Graphics Laboratory, http://gl.ict.usc.edu (accessed May 14, 2015).

feeling—i.e. the lack of microdetails—to the *nature* of the feeling itself. Why should an increase in human likeness to which there corresponds a decrease in familiarity—the uncanny valley—produce a sensation of aversion and disgust that makes our interaction with androids so problematic? We feel that a satisfactory answer to this question should ultimately come from philosophical analysis. In fact, we would like to suggest that the uncanny poses a deep philosophical challenge, one that we find appropriate to call *the problem of recognition*, and which manifests itself in an apparently deep-seated resistance to recognize characters falling in the uncanny valley as humans, to grant them a fully human status. Insofar as it crucially involves the highly elusive, and extremely hard to pin down notion of *humanness* or *human nature*, we believe that the problem of recognition should be seen and treated as mainly philosophical in nature. To the extent that we can broadly characterize *recognition* as a willingness to interact with other beings in an empathic way, the more or less intense feeling of revulsion typically experienced while visually inspecting characters which fall in the uncanny valley stands clearly in the way of that interaction. Yet, as already noticed, the very existence of the valley shows that the relation holding between recognition and pictorial realism is far more complex than one may initially suspect. Our willingness to engage in such interactions, in other words, does not seem to depend solely on the level of realism enjoyed by the characters we are supposed to interact with.

Before illustrating our working hypothesis, let us go back to Mori's seminal graph (see footnote 6). The shape of the curve which, according to Mori, captures the uncanny in quantitative terms has recently been questioned. The experimental data gathered by Bartneck et al. (2007) and Tinwell et al. (2011), for instance, seem to respectively support the existence of an uncanny *wall* or of an uncanny *cliff*, rather than an uncanny *valley*.[4] In general, most of the revisions so far proposed to Mori's graph have focused on the right hand side of the curve, whereas the left hand side has remained fairly noncontroversial. As a consequence, the left hand side of the curve seems to us better suited to develop qualitative considerations on the nature of the uncanny, and on the problem of recognition it poses.

Over the years, this problem has been approached from a wide variety of theoretical perspectives. MacDorman and Ishiguro (2006) have provided a useful map of the territory. A quick look at the main explanations they singled out may help us appreciate the complexity of our phenomenon. Here they are:

---

4   The three expressions (a) uncanny valley, (b) uncanny wall, and (c) uncanny cliff refer to three different shapes of the right hand side of the curve. See: Bartneck et al. (2007), Tinwell et al. (2011) and Tinwell, Grimshaw, and Williams (2011).

Table 7.1   Possible explanations of the Uncanny Valley

(A)	*Expectation Violation*	"The more humanlike the robot, the more human directed (largely subconscious) expectations are elicited. The fact that androids are often incapable of satisfying these expectations may be one reason why we perceive them to be not fully alive." (MacDorman and Ishiguro 2006, p. 309)
(B)	*Paradoxes Involving Personal and Human Identity*	According to Ramey (2005) an uncanny valley may result from "any cognitive act that links qualitatively different categories by quantitative metrics that call into question the originally differentiated categories. This effect can be especially pronounced when one of those categories is one's self or one's humanity. From a phenomenological standpoint, humanlike robots may force one to confront one's own being by creating intermediate conceptualizations that are neither human nor robot." (Ramey 2005, quoted in MacDorman and Ishiguro 2006, p. 310)
(C)	*Evolutionary Aesthetics*	"Another possible explanation for the uncanny valley is that androids are uncanny to the extent that they deviate from norms of physical beauty." (MacDorman and Ishiguro 2006, p. 310)
(D)	*Theory of Disgust*	"The natural defense mechanism of disgust may also be related to the uncanny valley." (ibid., p. 312)
(E)	*Terror Management*	"One hypothesis is that an uncanny robot elicits an innate fear of death and culturally-supported defenses for coping with death's inevitability." (ibid., p. 313)

Insofar as each and everyone of the above explanations seems to capture a different aspect of the uncanny, we believe that they should not be regarded as mutually exclusive. On the contrary, we believe that explanations A-E should all be seen as part of one and the same explanation of the phenomenon under investigation.

We are now ready to illustrate our working hypothesis, which consists in the following claim: Insofar as the emotion of *disgust* is connected to a process of *dehumanization*, this last notion may shed new light on our understanding of the uncanny. As stated, this hypothesis turns on two interconnected notions: the notion of *disgust*, on the one hand, and the notion of *dehumanization*, on the other. Let us briefly consider both in turn.

In 1987 Paul Rozin and April E. Fallon inaugurated a new season of research on disgust, which led to a more careful investigation of this emotion as well as to a deeper appreciation of its crucial role in human social and moral interactions. From an evolutionary perspective, disgust seems connected to the protection of the body from external threats. The well known physical reactions, such as nausea, typically associated with this emotion, are indeed meant to protect organisms from pathogens. In this regard, one of the merits of this new season of research was that of highlighting an important, although not yet fully understood, link between physical disgust, intended as a purely physiological reaction to various contaminants, and moral disgust, intended as a state of intellectual repugnance (see e.g. Rozin and Fallon 1987, Phillips et al. 1997, Rozin et al. 1999, Rozin, Haidt, and McCauley 2008, Rozin, Haidt, and Fincher 2009, Nussbaum 2004, 2010, Moll et al. 2005, Pizarro, Detweiler-Bedell, and Bloom 2006, Olatunji et al. 2007, Hodson and Costello 2007, Sherman and Haidt 2011, Buckels and Trapnell 2013). Indeed, according to this literature, the bridge between the two senses of the word, having been shaped by evolutionary processes, would be more than just metaphorical (Chapman et al. 2009, Rossi 2013). To put it in the vivid words of Rozin Haidt and Fincher (2009), we could describe this oral to moral evolution of disgust by saying that: "a mouth- and food-oriented rejection mechanism, a "get-this-out-of-my-body" emotion, has been elaborated (culturally and biologically) into a broad and meaning-rich emotion that protects not just the body but also the soul" (Rozin, Haidt, and McCauley 2008, p. 24).

The role played by the emotion of disgust in interpersonal relations has already been taken into serious consideration while reflecting on the nature of the uncanny, in that it clearly stands in the way of recognition, construed as a willingness to interact with other beings in an empathic way (MacDorman and Ishiguro 2006, Misselhorn 2009). This recognition failure, we believe, is best explained by supposing that the emotion of disgust is connected to a process of dehumanization.

Broadly speaking, the term "dehumanization" refers to the more or less conscious and more or less intentional denial of either humanness *tout court*, or single human traits to other members of our species. This denial typically manifests itself in private and social behaviors aimed at discouraging or punishing social interactions with stigmatized individuals or groups. From the religious wars described in the Bible to various forms of colonialism, the recorded history of our species has made episodes of dehumanization sadly familiar. On a cognitive level, behaviors linked to dehumanization are typically underpinned by the emotion of disgust, broadly construed (Rozin and Fallon 1987, Rozin et al. 1999, Rozin, Haidt, and McCauley 2008, Rozin, Haidt, and Fincher 2009, Haslam 2006, Haslam et al. 2008). In a recent survey of the many theoretical uses of this notion, Nick Haslam has distinguished two main kinds of dehumanization, to which there would correspond two distinct notions of humanity (2006). What Haslam calls *animalistic dehumanization* consists in the denial to other human beings of traits that are perceived as uniquely human, among which he lists civility, refinement, moral sensibility, rationality, and maturity (as opposed to childlikeness). What he calls *mechanistic dehumanization*, on the other hand, is perpetrated by denying other humans of traits that are perceived as somehow constitutive of the human nature, among which he includes emotional responsiveness, interpersonal warmth, cognitive openness, agency, and depth (as opposed to superficiality). As Haslam's labels suggest, whereas in the first case we tend to think of our human other as an *animal*, in the second we rather think of it as a *machine*. We believe that holding the uncanny against the backdrop of this distinction may shed new light on its nature. Indeed, there seems to be a clear sense in which a process of dehumanization is at work in our instinctive and largely emotional reactions to androids or photoreal digital characters.

The core of our hypothesis, in a nutshell, is that the notion of dehumanization, while certainly in need of more careful clarification, represents an apparently promising and yet so far virtually unexplored conceptual bridge between the notion of disgust, on the one hand, and the notion of uncanny, on the other. In particular, we believe that the complex phenomenon represented by the left hand side of Mori's graph, which all of the explanations A-E above try to account for in one way or another, could be usefully located at the intersection between Haslam's two notions of animalistic vs. mechanistic dehumanization. As a consequence, we believe that by exploring this intersection area one could reach precious insights into the nature of the uncanny.

References

Alexander, O., G. Fyffe, J. Busch, X. Yu, R. Ichikari, A. Jones, P. Debevec, J. Jimenez, E. Danvoye, B. Antionazzi, M. Eheler, Z. Kysela, and J. von der Pahlen. 2013. "Digital Ira: Creating a Real-Time Photoreal Digital Actor." ACM SIGGRAPH 2013 Posters, Anaheim, California. doi: http://dx.doi.org/10.1145/2503385.2503387.

Alexander, O., M. Rogers, W. Lambeth, M. Chiang, and P. Debevec. 2009. "The Digital Emily Project: Photoreal Facial Modeling and Animation." ACM SIGGRAPH 2009 Courses, New Orleans, Louisiana. doi: http://dx.doi.org/10.1145/1667239.1667251.

Angelucci, A., M. Bastioni, P. Graziani, and M.G. Rossi. 2014. "A Philosophical Look at the Uncanny Valley." In *Sociable Robots and the Future of Social Relations: Proceedings of Robo-Philosophy 2014*, edited by Johanna Seibt, Raul Hakli and Marco Nørskov, 165-171. Amsterdam: IOS Press Ebooks. doi: http://dx.doi.org/10.3233/978-1-61499-480-0-165.

Bartneck, C., T. Kanda, H. Ishiguro, and N. Hagita. 2007. "Is the Uncanny Valley an Uncanny Cliff?", Robot and Human interactive Communication, 2007. RO-MAN 2007. The 16th IEEE International Symposium on, 26-29 Aug. 2007. 368-73. doi: http://dx.doi.org/10.1109/ROMAN.2007.4415111.

Bastioni, M. and P. Graziani. 2011. "Quanta Matematica Serve Per Costruire Un Essere Umano?" *Lettera Matematica PRISTEM* 78: 26-37.

Buckels, E.E. and P.D. Trapnell. 2013. "Disgust Facilitates Outgroup Dehumanization." *Group Processes & Intergroup Relations* 16(6): 771-80. doi: http://dx.doi.org/10.1177/1368430212471738.

Chapman, H.A., D.A. Kim, J.M. Susskind, and A.K. Anderson. 2009. "In Bad Taste: Evidence for the Oral Origins of Moral Disgust." *Science* 323(5918): 1222-1226. doi: http://dx.doi.org/10.1126/science.1165565.

Freud, S. 1953. "The 'Uncanny'." In *The Standard Edition of the Compelete Psychological Works of Sigmund Freud*, edited by James Strachey, 219-52. London: The Hogarth Press.

Gouskos, C. 2006. "The Depths of the Uncanny Valley." CBS Interactive Inc Accessed April 21, 2015. http://www.gamespot.com/articles/the-depths-of-the-uncanny-valley/1100-6153667/.

Haslam, N. 2006. "Dehumanization: An Integrative Review." *Personality and Social Psychology Review* 10(3): 252-64. doi: http://dx.doi.org/10.1207/s15327957pspr1003_4.

Haslam, N., Y. Kashima, S. Loughnan, J. Shi, and C. Suitner. 2008. "Subhuman, Inhuman, and Superhuman: Contrasting Humans with Nonhumans in Three Cultures." *Social Cognition* 26(2): 248-58. doi: http://dx.doi.org/10.1521/soco.2008.26.2.248.

Hodson, G. and K. Costello. 2007. "Interpersonal Disgust, Ideological Orientations, and Dehumanization as Predictors of Intergroup Attitudes." *Psychological Science* 18(8): 691-8. doi: http://dx.doi.org/10.1111/j.1467-9280.2007.01962.x.

MacDorman, K.F. 2005. "Androids as Experimental Apparatus: Why Is There an Uncanny Valley and Can We Exploit It?" *CogSci-2005 Workshop: Toward Social Mechanisms of Android Science*, CogSci-2005 Workshop, Stresa, Italy, July 25-6. 108-18.

MacDorman, K.F., R.D. Green, C-C. Ho, and C.T. Koch. 2009. "Too Real for Comfort? Uncanny Responses to Computer Generated Faces." *Computers in Human Behavior* 25(3): 695-710. doi: http://dx.doi.org/10.1016/j.chb.2008.12.026.

MacDorman, K.F. and H. Ishiguro. 2006. "The Uncanny Advantage of Using Androids in Cognitive and Social Science Research." *Interaction Studies* 7(3): 297-337. doi: http://dx.doi.org/10.1075/is.7.3.03mac.

MacDorman, K.F., S.K. Vasudevan, and C-C. Ho. 2009. "Does Japan Really Have Robot Mania? Comparing Attitudes by Implicit and Explicit Measures." *AI & Society* 23(4): 485-510. doi: http://dx.doi.org/10.1007/s00146-008-0181-2.

Misselhorn, C. 2009. "Empathy with Inanimate Objects and the Uncanny Valley." *Minds and Machines* 19(3): 345-59. doi: http://dx.doi.org/10.1007/s11023-009-9158-2.

Moll, J., R. de Oliveira-Souza, F.T. Moll, F.A. Ignácio, I.E. Bramati, E.M. Caparelli-Dáquer, and P.J. Eslinger. 2005. "The Moral Affiliations of Disgust: A Functional Mri Study." *Cognitive and Behavioral Neurology* 18(1): 68-78.

Mori, M., K.F. MacDorman, and N. Kageki. 2012. "The Uncanny Valley." *IEEE Robotics & Automation Magazine* 19(2): 98-100. Original edition, 1970. doi: http://dx.doi.org/10.1109/MRA.2012.2192811.

Nussbaum, M.C. 2004. *Hiding from Humanity: Disgust, Shame, and the Law.* Princeton, NJ: Princeton University Press.

Nussbaum, M.C. 2010. *From Disgust to Humanity: Sexual Orientation and Constitutional Law, Inalienable Rights.* Oxford: Oxford University Press.

Olatunji, B.O., N.L. Williams, D.F. Tolin, J.S. Abramowitz, C.N. Sawchuk, J.M. Lohr, and L.S. Elwood. 2007. "The Disgust Scale: Item Analysis, Factor Structure, and Suggestions for Refinement." *Psychological Assessment* 19(3): 281-97. doi: http://dx.doi.org/10.1037/1040-3590.19.3.281.

Phillips, M.L., A.W. Young, C. Senior, M. Brammer, C. Andrew, A.J. Calder, E.T. Bullmore, D.I. Perrett, D. Rowland, S.C.R. Williams, J.A. Gray, and A.S. David. 1997. "A Specific Neural Substrate for Perceiving Facial Expressions of Disgust." *Nature* 389(6650): 495-8. doi: http://dx.doi.org/10.1038/39051.

Pizarro, D.A., B. Detweiler-Bedell, and P. Bloom. 2006. "The Creativity of Everyday Moral Reasoning: Empathy, Disgust, and Moral Persuasion Creativity and Reason in Cognitive Development." In *Creativity and Reason in Cognitive Development*, edited by James C Kaufman and John Baer, 81-98. Cambridge: Cambridge University Press.

Plantec, P. 2008. "The Digital Eye: Image Metrics Attempts to Leap the Uncanny Valley." *VFXWorld Magazine*, August 7, 2008. Accessed May 14, 2015. http://www.awn.com/vfxworld/digital-eye-image-metrics-attempts-leap-uncanny-valley.

Ramey, C.H. 2005. "The Uncanny Valley of Similarities Concerning Abortion, Baldness, Heaps of Sand, and Humanlike Robots." *Proceedings of the Views of the Uncanny Valley Sorkshop*, 2005 IEEE-RAS International Conference on Humanoid Robots Tsukuba, Japan, December 5.

Rossi, M.G. 2013. *Il Giudizio Del Sentimento: Emozioni, Giudizi Morali, Natura Umana.* Rome: Editori Riuniti University Press.

Rozin, P. and A.E. Fallon. 1987. "A Perspective on Disgust." *Psychological Review* 94(1): 23-41. doi: http://dx.doi.org/10.1037/0033-295X.94.1.23.

Rozin, P., J. Haidt, and K. Fincher. 2009. "From Oral to Moral." *Science* 323(5918): 1179-80. doi: http://dx.doi.org/10.1126/science.1170492.

Rozin, P., J. Haidt, and C.R. McCauley. 2008. "Disgust: The Body and Soul Emotion in the 21st Century." In *Disgust and Its Disorders*, edited by B O Olatunji and D McKay, 9-29. Washington, DC: American Psychological Association. https://sites.sas.upenn.edu/sites/default/files/rozin/files/260disgust21centolatunji2008.pdf.

Rozin, P., L. Lowery, S. Imada, and J. Haidt. 1999. "The CAD Triad Hypothesis: A Mapping between Three Moral Emotions (Contempt, Anger, Disgust) and Three Moral Codes (Community, Autonomy, Divinity)." *Journal of Personality and Social Psychology* 76(4): 574-86. doi: http://dx.doi.org/10.1037/0022-3514.76.4.574.

Sherman, G.D. and J. Haidt. 2011. "Cuteness and Disgust: The Humanizing and Dehumanizing Effects of Emotion." *Emotion Review* 3(3): 245-51. doi: http://dx.doi.org/10.1177/1754073911402396.

The Economist. 2010. "Crossing the Uncanny Valley." *The Economist*, November 18, 2010. Accessed May 14, 2015. http://www.economist.com/node/17519716.

Tinwell, A. 2014. "Applying Psychological Plausibility to the Uncanny Valley Phenomenon." In *Oxford Handbook of Virtuality*, edited by M Grimshaw, 173-86. Oxford: Oxford University Press.

Tinwell, A., M. Grimshaw, D.A. Nabi, and A. Williams. 2011. "Facial Expression of Emotion and Perception of the Uncanny Valley in Virtual Characters." *Computers in Human Behavior* 27(2): 741-9. doi: http://dx.doi.org/10.1016/j.chb.2010.10.018.

Tinwell, A., M. Grimshaw, and A. Williams. 2011. "The Uncanny Wall." *International Journal of Arts and Technology* 4(3): 326-41.

Chapter 8

Staging Lies: Performativity in the Human–Robot Theatre play *I, Worker*

Gunhild Borggreen

This chapter takes the Japanese Robot–Human Theatre production *Hataraku Watashi* (*I, Worker*) by Hirata Oriza and Ishiguro Hiroshi from 2008 as its starting point. An analysis of the theater play leads to an investigation of the complex layers of intention, effect, and cultural conventions in human–robot interaction. Linking the fiction of the stage production to laboratory testing that include robots in simulated real-life situations, the notion of how robots "lie" are discussed in terms of different types of participants: from Wizard-of-Oz methods in HRI (human–robot interaction) research to the theories of performative speech acts from the British philosopher of language J.L. Austin's *How To Do Things With Words*. The text includes discussions of how gender and other social norms guide the imagination of future technologies. The text discusses how robots may act "parasitic" upon normal circumstances and includes theories of iteration and performativity by Jacques Derrida and Judith Butler to point out the troubled notions of "normal." Iterative processes provide a potential for agency in robot–human interaction, and emphasize the importance of aesthetics as a framework of reflection in robotics research.

Introduction

During a research stay in Japan in the spring of 2011, I had the opportunity to watch a live performance of the theater play *Hataraku watashi* (I, Worker), a robot–human theater production from 2008, written and directed by theater director Hirata Oriza in collaboration with robot scientist Ishiguro Hiroshi. The play was staged as part of the program of Tokyo Performing Arts Meeting (TPAM), an annual event in which the latest productions from the Japanese theater and performance art scene are presented to an audience of domestic and international scholars, theater producers and general public. The TPAM featured a special section on robot–human theater that year, and staged another of Hirata and Ishiguro's collaborations entitled *Sayonara* (Goodbye, from 2010). The event also included a presentation by Nomura Masashi, the manager of the Tokyo-based Seinendan Theatre Company behind the two robot–human theater plays. Nomura's presentation focused on the dual purpose of staging a robot–human theater play:

it is a performance, but at the same time it is also a social experiment because the play will be used for what Nomura calls "next praxis." As he stated: "while at this moment people see this performance from the audience seats, they are destined to be performers on the stage in the near future"(TPAM Direction 2011).

The narrative plot in the play *I, Worker* displays examples of how robots and human beings are imagined to interact and communicate in a not so distant future. In this sense, Nomura may be correct to assume that what is presented on the stage will be part of ordinary peoples' everyday life some years from now. In order to utilize the play as an imaginary "window" into future societies, audiences were given a questionnaire as they left the theater hall afterwards, and asked to answer questions such as whether is was easy to distinguish the two robot from each other, or to which degree each of the robots were likely to understand the mood of the other characters. Surely, the play was an aesthetic manifestation of Hirata's and Ishiguro's imaginations concerning likely scenarios of robot–human interaction, but the play also contributed to scientific investigations related to social robotics. So one important question to pose to an analytical perspective of the play *I, Worker* would be to focus on the borderline between "real life" and art in terms of robotics and ask: what can artists (such as Hirata) and robot engineers (such as Ishiguro) learn from each other? Why is robotic art and aesthetics important to the development of science and technology, and how can this play and other artistic formats contribute to the inquiry of ontologies that seems to be crucial for the field of social robotics?

In a scene in the play *I, Worker*, one of the robots declares: "robots cannot lie." The words are conveyed by the robot in the middle of a conversation between two robots and a human being, and the remark is not emphasized in any particular way in the play text. However, as I will argue in the following, it appears to be a key moment in the text because the remark may function as a kind of ontological litmus test: if robots in fact *can* lie, then the remark itself is a lie. On the other hand, if it is true that robots cannot lie, the particular property of lying may be one of the ways in which to distinguish humans from robots. As elaboration of previous analyses, I focus here on the concept of lying and use it as my entrance into an analysis of the play *I, Worker* followed by a discussion of the aesthetics of theater and the concept of performativity (Borggreen 2014).

Robots in the Family

The play *I, Worker* focuses on dialogues and subtle interactions between human beings and robots. The play features a young couple Yūji and Ikue, and the narrative takes place in the living room of their home. Yūji and Ikue have no children, but they have two humanoid robotic home companions to help out in the house: Takeo, a male-gendered robot, and Momoko, a female-gendered robot, wearing an apron during the play. The roles of Yūji and Ikue were performed by two human actors, while the two robots were performed by two bright yellow Wakamaru robots from Mitsubishi Heavy Industry. Each robot is 1m tall, they are mounted on wheels, and they can move their arms and rotate their heads and bodies. The robots speak Japanese mediated in synthetic-sounding voices, in contrast to the natural human voices from the human actors. The play is a series of everyday situations where the humans and the robots have interaction and communication, and may be inspired by the many scenarios included in social robotics. Although still on an experimental stage, a lot of research in Japan focuses on how to develop humanoid robots that can function as partners and helpers for human beings in everyday life. Due to the demographic challenge in Japanese society in terms of an aging population and a decreasing birthrate, the government has created plans of implementing social robots in various institutions in society, such as schools and hospitals, and more importantly, to implement robots in the home and within the family. In 2008, the Japanese Cabinet Office presented a plan called Innovation 25, which sketched out a number of goals to be carried out before 2025, many of which included robots. Issue number 9, for example, proposes "one robot in each home" (*ikka ni ichidai katei robotto*) on lease basis as a part of the new service economy. The household robot is imagined to carry out household tasks such as cleaning and doing the laundry, and hereby create more time for child-rearing, work, and hobbies. The household robot will also be able to support elderly or sick people who want to stay in their own home, and need everyday assistance (Inobeeshon 25 2008).

Anthropologist Jennifer Robertson (2014) has studied some of the many dimensions of household robots in Japan, and elaborates on the possibility that humanoid robots may in the future even become legitimate members of the household and attain citizenship as part of the *koseki*, a registry of the extended family household in Japan known as *ie*. According to Robertson, the traditional *koseki* system sustains gender roles and family structures and in extension supports the ethnic homogeneity of Japan by placing emphasis on blood lineage and decent. At the same time, nationwide surveys conducted by Cabinet Office seem to show that many Japanese people are uncomfortable with the idea of having foreign workers to tend to them when sick or in need of care due to old age, so the Cabinet Office's Innovation 25 plan for household robots fits nicely into this apparent demand in the health care sector. Exactly to which degree such grandiose plans for the implementation of robots as companions and household helpers in every Japanese home will be carried out and function in reality, is too early to predict: the limited physical space in average Japanese houses and thus the restrictions to actually accommodate a robotic entity moving about in the home is just one of countless challenges that lie ahead.

In other aspects of her studies of robots in Japan, Jennifer Robertson (2007, 2010) discusses gender roles and investigates the relationship between robots and fertility. One of the reasons for the shrinking Japanese population is often claimed to be young women's reluctance to marry and have children, so according to Robertson, one of the goals of the Cabinet Office's plans for the implementation of robots is to make it more attractive for young women to become housewives and mothers because technology can help out in the house. The birth rate in Japan being as low as 1.3 children per married woman, and scholars see this as a protest against a social system in Japan that does not offer equal opportunities for men and women, but continues to regard women as "second-class citizens" (Robertson 2010, p. 10). In other words: the Japanese government, probably heavily supported by the Japanese robot industry, seems to be considering a technical solution to a problem that has much wider social, political, and economic implications. The particular attention of women as the future users of robot technologies in the home show how new technologies are often imagined and developed within already existing frameworks of social and cultural practices, and that social robotics thus is an integral part of highly ideological field such as ethnicity, gender roles, or national identity. This aligns with the Robertson's conclusions on her study of the implementation of social robots in Japan, namely that traditional social values and structures are reinforced through the implementation of robotic technologies in care-taking industry and entertainment.

Identity Crisis

Indeed, the gender roles and the nuclear family structure represented in the play *I, Worker* follow the conventional ideals in Japanese society. The conversations taking place in the play are based on everyday exchange among people (and in this case also robots), and the dialogues are rendered in an informal and colloquial manner. Sentences are short, and not overtly emotional, but reveals as a subtext various patterns of co-existence and knowledge about each other that the humans and robots in the play appear to have developed among each other over time. Parts of the conversation between the robots and the human beings in *I, Worker* reveal various aspects of identity crisis, both on parts of the humans as well as for the robots, and there are numerous small incidents of subtle emotions and a lot of apologies. The issue of fertility and child bearing is addressed in the play, albeit in an indirect manner. At some point, the female robot Momoko has a conversation with female human being Ikue, and refers to the initial stage of robot development, where robot technology was still on a primitive level, the most difficult thing was to hold an egg without breaking it. This has changed: "now," the robot continues, "we can even hold human babies." After a pause in the conversation, the robot apologizes for bringing up the topic of children, referring to that fact that Ikue and Yūji do not have any. Ikue replies: "It's okay, you don't have to empathize with that. I suppose there are ways I could try to have one," but she also expresses concern about the future of possible child (Hirata 2010, p. 44). This kind of anxiety about the future of new generations may well reflect some of the concerns on which young women in the current Japanese society base their decisions for not wanting children, and the conversation in the play also reflects the tension when others (in the family, among friends) constantly bring up the issue of children as a mild and indirect form of pressure.

Another example of crisis concerns the male robot Takeo, who does not want to leave the house to help Ikue go shopping. In the robot's self-understanding, this aversion to go outside in order to help the human being is a major crisis because it contradicts the very purpose of being a robot. As the title *I, Worker* (The original title in Japanese *Hataraku watashi* may be translated to "I who work") suggests, parts of the crises evolve around how having a job is part of identity formation. The robot Takeo states: "a robot should be working, we are created to work." The "identity crises" of the robots mirror those of the humans. It turns out that Yūji, the male human being, is out of job and also does not want to leave the house. This poses a problem for his wife Ikue in her relationship to her parents. In Ikue's conversation with the robot Takeo, she explains how her parents keep bothering Yūji that he has no job: "My parents say that to Yūji all the time. 'People are meant to work.' 'You have to be able to feed yourself through hard work'" (Hirata 2010, p. 41). In this way, the dialogue on stage poses questions concerning issues such as gender roles, family values and social expectations, which reflects real-life issues that many Japanese face in the post-bobble era of economic recession.

The surprising effect of the play is of course when emotions related to existentialist concerns are expressed not by a human being but by a machine. This *Verfremdung* effect created by the juxtaposition of human emotions expressed through machines is constantly and explicitly addressed in the play, and becomes the key to questioning the possible differences and similarities between humans and robots. This is where the remark about lying may be crucial. In the narrative of the play, the human beings take the emotional response from the robots rather seriously. In a scene following the "identity crisis", the male human Yūji tries to cheer up the miserable robot Takeo by playing the theme song from RoboCop on his CD player. But then Yūji realizes that playing a heroic battle tune was inappropriate and he apologizes to the robot. He bows his head and says "I'm sorry about before. I thought it would cheer you up." The robot replies by dismissing the incident: "It is nothing. It is not your fault," the robots says. The other robot standing next to them adds, as a way of confirming the truth in this remark: "robots cannot lie" (Hirata 2010, p. 49).

Lying

"What is a lie?" asks social anthropologist J.A. Barnes in his book *A pack of lies. Towards a sociology of lying* from 1994. According to Barnes, lies, falsehood, and deceptions have been part of human culture since ancient Greek or even before, and appears to be an important part of social interaction. Most societies have religious and ethic codes against lying, and it appears that a preference for the truth is ubiquitous, if not universal. In many cultures people teach their children not to lie, but lying is nevertheless an important part of growing up because a child's first lie is seen by psychiatrists as "a decisive further step into separateness and autonomy" (ibid., p. 8). Philosophers see lying as part of the foundation of subjectivity, and Barnes argues that the attraction of compiling lies is due to "the perception of lying as evidence for freedom and imagination." Barnes also refers to Thomas Hobbes and other philosophers who have emphasized "the ability to lie as one of the criteria that distinguish human beings from other animals" (ibid., p. 3). Studies of nonhuman primates, however, show that animals are capable of some sorts of intentional deception by using their voices to deceit other animals, or practice other modes of lying. Sociobiologists and ethologists study various types of deception, and some suggest a typology for deception with four levels that increase in complexity: deception by appearance, deception by actions, learned deception, and planned deception. Even the most complex level includes animals such as chimpanzees and baboons in addition to, as Barnes laconically remarks, "innumerable humans." In other words, it is not possible to use the notion of lying as a tool to distinguish humans from other animals. But what about robots?

Numerous examples of the testing of robots within science and technology are based on deception. The method known as the Wizard-of-Oz (WoZ) is often applied in the field of HRI, human–robot interaction. The term is a reference to the character Oz in the book *The Wonderful Wizard of Oz* by L. Frank Baum from 1900, a popular children's novel which was adapted for film in 1939. In most of the novel, the Wizard remains unseen, and when he does appear, it is in disguise, thus lending the notion of concealment to this particular robotic research method. The Wizard-of-Oz method includes a person who operates the robot remotely, usually hidden from the test person who interacts with the robot. As roboticist Laurel D. Riek (2012) notes in her systematic survey of a number of WoZ experiments in HRI, the Wizard-of-Oz method may be used in cases where the robots are not sufficiently advanced to interact autonomously with humans, or as a test of the robot at early stages of an interaction design process. WoZ methods are most commonly applied for natural language processing and non-verbal behavior simulation, while the hidden wizard may also perform navigation and mobility tasks with the robot. Riek concludes that the use of Wizard-of-Oz methods are likely to continue in the future because the most challenging element of robot development in HRI is that of verbal and visual human language understanding. The WoZ method of puppeteering the robot allows people (test participants as well as the scientists) to envision what human–robot interaction may be like in the future. However, as Riek notes, the robot in this kind of remote controlled puppeteering serves more as a proxy for humans and less as an independent entity. In fact, the situation may be described not as a human–robot interaction, but as a human–human interaction via a robot.

Theatre-based Human–Robot Interaction

Because the Wizard-of-Oz method includes staging fiction-like scenarios, it may be easy to understand how the notion of theater aligns well with the study of robot–human interaction. Hence, a number of HRI research projects apply theater as a methodology. In 2003, roboticist Cynthea Breazeal and her colleagues at MIT Media Lab published a paper on Interactive Robot Theatre, in which they report on their experience with a specific performance they set up (Breazeal et al. 2003). In the development of robotic assistants that cooperate with people as partners rather than merely a tool, it is important to develop robots that can "interact naturally and appropriately with people" (Breazeal et al. 2003, p. 78). The authors do not explicate what "natural" and "appropriate" interaction is, but they note that faces, gesture, and speech are "the natural interfaces people use to communicate with one another", and therefore sociable robots should be able to perceive, recognize, and interpret human behavior through multiple modalities including vision, audition, and touch. Interactions are always unpredictable, and therefore impossible to integrate in the controlled environment of a robot laboratory. Live theatrical performances may function as a testing lab for robots' ability to navigate in uncertain terrain because live performance, according to Breazeal and her colleagues, will have some stabile elements, such as storyline and stage design, while other elements are unknown. This semi-controlled scenario, they argue, may help the robot engineers and designers to create robots more attuned to the core challenges of human–robot interaction, namely the robot's ability to perceive, interpret, and respond to the open-ended and unpredictable behavior of human beings. The elements that make up the unpredictability will appear on stage (and thus "challenge" the robot) as "improvisation" or "audience participation," elements which characterizes the "good" actor's ability to not just "act" but also to "react" in "a convincing and compelling manner" in relation to the performance of others.

Another and different example is LIREC, an integrated project based at University of Hertfordshire in which researchers use a Theatre-based Human–Robot Interaction methodology (THRI). One article published in the context of LIREC states that theater methodology can be seen as a means of prototyping, and that it provides a "sense of physical presence and real-time performance" so essential for HRI (Syrdal et al. 2011). In another test situation of the LIREC project, an actor plays the role of a robot owner in front of a group of volunteer test persons. The robot too is "acting" by remote control because it is made to simulate actions that no robot can actually perform yet, such as robust navigation in everyday environments, smooth natural language interaction, and "social intelligence" in interaction and dialogue with the human owner (Chatley et al. 2010). The scenarios with the actor and the robot in the LIREC project includes extensive use of the Wizard-of-Oz method. In the left hand side of the room and off-stage, technicians operate the robot without the audiences' knowledge. According to the research report, the actor who pretends to be the robot owner will stay in character during the short scenarios and during the discussions afterwards in order to "make the audience believe that the experiment was a true representation of an owner and robot". Likewise, those technicians who control the robot are placed "out of the audience's direct line of sight to allow the audience to 'forget they are there'" (ibid., pp. 74-5). This is an example of how scientific testing of human–robot relations apply performance as a means of producing knowledge about human response. The authors focus on realism as an aesthetic modus when they describe their approach as "a theatrical presentation with an actor interacting and cooperating with robots in realistic scenarios before an audience" (ibid., p. 73). According to other articles on human–robot interaction as theatre, it appears to be commonly thought that actors and robots have the goal to "get as close as possible to the unobtainable ideal", which is presenting the actual character to the audience. In this process, "the robot/actor must appear to be something it is not." Hence, many HRI researchers, for example David V. Lu and William D. Smart (2011), think of actors and robots on equal terms and argue that the theatrical stage for the human actor is equivalent to the real-life scenarios for the social robot because both situations rely on the "*as-if*" effect.

Performatives and Infelicities

The "as-if" effect is a deception or a lie that makes up the core of these types of theatrical settings with robots: the interaction between robots and human appear to be authentic, but is in fact scripted and rehearsed. This goes for the HRI theater model in the LIREC project as well as the play *I, Worker*. There is, however, a difference in the way in which the lie is staged when comparing robots appearing on stage in a theater production and the theatre-like scenarios for testing human–robot interaction. In order to analyze this difference I will turn to linguistic philosopher J.L. Austin and his theory on performative speech acts. In performance studies, Austin is credited for coining the term "performative", which denotes the kind of linguistic utterances that perform or "do" an action when they are spoken. Austin's theory of performative speech act was first given as a lecture series in the 1950s, and his notes were later published as *How To Do Things With Words* (1975). A performative sentence is not just simply to state something, but is to carry out an action through speech. One of Austin's examples that appear in his lectures is the utterance "I do" in a wedding ceremony: by uttering the words under the proper circumstances a conjugal union is established. Discussing the properties of performatives, Austin is less concerned about whether an utterance is true or false; instead he focuses on the success or failure of an utterance, and creates a distinction between "felicitous" and "infelicitous" utterances. The utterance "I do" may thus only be successful or felicitous if it fulfills certain conventions, for example if it is spoken by a person involved in a wedding ceremony, and by a person who is not already married. Austin's performatives are relevant for exploring social interaction because he is concerned about intentionality (of the speaker) and effect (on the listener) in everyday situations of linguistic communication. Austin's linguistic theory is therefore often linked with anthropological and social performance through semiotics and the notions of social and cultural conventions (Carlson 1996).

In his argument, Austin presents a number of infelicities, and one of them concerns situations where there is no true intention behind the utterance. This happens, according to Austin, when a performative is uttered on a theater stage or in poetry. Austin states: "a performative utterance will, for example, be in a peculiar way hollow or void if said by an actor on the stage, or if introduced in a poem" (1975, p. 22). Language in such circumstances, Austin continues, "is in special ways—intelligibly—used not seriously, but in ways parasitic upon its normal use." For Austin, "happy" or smooth performatives are those spoken in ordinary everyday circumstances, and there must exist "an accepted conventional procedure having a certain conventional effect." He dismisses performative utterances that are spoken in what he calls "unserious" ways, that: in aesthetic or fictional situations such as theater and poetry. Austin considers words spoken on a stage to be "infelicitous" because there is no true intention behind.

Within the field of performance studies, this particular statement by Austin has been the core of ongoing discussions about the aesthetics of theater and performance. Despite the innovative creation of new concepts such as the performative, Austin has also been seen as somewhat misguided in terms of the characteristics of stage art. Performance scholar Richard Schechner, for example, suggests that Austin did not understand "the unique power of the theatrical as imagination made flesh" (2006, p. 124). According to Schechner, Austin did not appreciate the way in which a performative utterance on stage creates a liminal, transitional, and intermediary space; it creates another type of reality which is different from everyday reality, but not its opposite. Other performance scholars have tried to elaborate, explore, and restore what Austin meant with the remark upon a "parasitic" use of language on stage. Performance scholar Branislav Jakovljevic (2002), for example, suggests that by considering the entire theatrical act, and not just the performative utterance spoken on the stage, the act can achieve an impact similar to, and not as a paradox to, Austin's performative speech act. The issue about Austin's "unserious" deception and "parasitic" usage of language in poetry or on stage is related to aesthetic framing and the way in which "reality" of the audiences relates to the fictitious environment on stage.

The Aesthetics of Robot Theatre

Comparing the two different types of robot–human theaters settings mentioned above, it will be clear that Austin's notion about the use "parasitic upon its normal use" takes different forms depending on the aesthetic framing of the event. It is exactly the aesthetic framing that makes the difference. In the case of the Hertfordshire test for robotic home companions, the test persons know that they are engaging in a scientific experiment, but they are *not* aware of the theatrical framing of the entire event. The test persons may very well think that robots are able to actually behave and speak as they do in the scenarios. Hirata Oriza's play *I, Worker* resembles the robot testing at Hertfordshire University: here too the robots are manipulated to act as if they can carry out robust navigation in everyday environments, engage in smooth natural language exchange, and display "social intelligence" in interaction and dialogue with the human owners, in this case Yūji and Ikue. There is, however, a significant difference in how these two types of staging are perceived because *I, Worker* takes place in a theatre, and has already before the play begins been announced and contextualized as a work of art. The audiences are not fooled in the same way as the test audience at the Hertfordshire laboratory, but instead engage knowingly and voluntarily in the "deception" of the theatrical framing, and they are concsious about the aesthetic choices and setting of the play.

Ishiguro Hiroshi, the robot scientist, explains the reasons for him to collaborate with Hirata Oriza in an essay on robot science entitled "Robotto kenkyū towa ningen no kokoro o shirukoto" ((To know the human heart through robot science) Ishiguro 2010). For Ishiguro, the naturalism in Hirata's play is the main reason for the collaboration because this is where the artistic and the scientific approach have central elements in common. One of Ishiguro's goals is to approach the notion of a human heart. "We robot scientists cannot programme a human heart into a robot", Ishiguro writes. "We can programme a function that makes it look *as if* the robot has a heart. But the problem is that we do not know what it looks like to have a heart." Ishiguro notes that psychologists and cognitive scientists study such aspects of human nature, but their experiments are always carried out in the controlled environments of the laboratory, and therefore "cannot clearly explain how humans express their heart in the midst of multiple stimuli in everyday situations" (ibid., pp. 55-6). When scientific approaches such as robotics, psychology, and cognitive science cannot give the answer, Ishiguro turns to the artistic creation of Hirata's theater work.

It may, however, be difficult for Ishiguro, the robotic scientist, to get any scientific answers to the questionnaire that was distributed to audiences at the *I, Worker* performance at TPAM in Yokohama or similar performances. A question such as if the respondent thought that the robot was able to understand the mood of other characters in the play is not easy to answer—not because of any real or imagined difference between human beings or robots, but because the aesthetics of the play itself. In any play by Hirata Oriza, even if all actors were human beings, it would be a challenge to give definite explanation of the characters' emotions and moods. Hirata's plays do not feature dramatic action; conversations are based on suggestive and indirect communication; the stage is often constructed in minimalist style with few or no references to specific places, and the play offers no clues to the audiences as to how to interpret the relations or emotions that take place among the characters. If it is difficult to characterize the mood of the robots in the play *I Worker*, it is equally difficult to characterize the mood of the human beings. The variety of answers that may come up will always be subjective, and will be based upon the personal experience and knowledge of the world that each of the audience brings into the theatrical space. Individual experience and knowledge is activated through an interaction between the performance and the viewer; but it will never be the same for two persons, and will never be able to repeat itself. This is what art can offer to science: no specific or concrete answers, but an open and dynamic space of imaginary visions that can be changed and negotiated.

Naturalism as a Set-Up

In the play *I, Worker* the relationship between reality and fiction is challenged by the aesthetic format of the play. *I, Worker* resembles a naturalistic play because it contains what performance studies scholar Peter Eckersall (2011) describes as "visceral conditions of life through corporeality and sensation and without moral intervention." Hirata Oriza and his Seinendan Theatre Company is acclaimed for the use of contemporary colloquial language, and many critics praise the way in which Hirata dissects ordinary everyday situations in his plays. As theater scholar M. Cody Poulton (Hirata and Poulton 2002) describes, Hirata's aesthetics is characterized as *shizuka na engeki* (quiet theatre), a contemporary development of the naturalistic *shingeki* or "new theatre" from the 1920s. Quiet theater refers to Hirata's reduction of speech, as opposed to the many Western plays that were translated and staged in Japanese in the conventional *shingeki* format. Hirata evented the quite speech mode because he found that the long, emotionally explicit dialogues in Western naturalist plays do not match the everyday colloquial speech. So the conversation in Hirata's play, including *I, Worker*, is seldom explicit about the characters' emotions and feelings, but something revealed in between the lines, as a subtext. There are long breaks and pauses in the dialogues, just as there are in real life conversations. Remarks may be laconic, some topics are avoided, there are no direct insight into the inner thoughts of the characters. The kind of return to the aesthetics of "the real" is manifest in *I, Worker* by the fact that one of the actors was present on stage already before the show had started, lying on the stage floor reading a magazine while audiences were entering the theater hall and finding their seats. The robots too would roll across the stage and disappear again once or twice, as if they were executing some task independent from what was going on in the theater space; as if the play had already begun without its audience. This type of dramaturgy is based on the desire to recover "the real" and privilege life "as it is" (*ari no mono*), and can according to Poulton be seen as Hirata's way of challenging the conventions of orthodox *shingeki* or naturalist theatre. This way of framing the entire theatrical setting, including the theater space itself, the audiences behavior, as well as the time before the play starts, may be similar to the way in which Jakovljevic refers to the entire theatrical act rather than the actual words spoken on stage as a means of negotiating Austin's notion of "parasitic" use of language in theatre.

The play *I, Worker* is different from other of Hirata's plays because the colloquial dialogues of everyday life here takes place not between human beings, but between humans and machines. The elements of improvisation and individual response usually associated with naturalistic theater is at stake in this play. The naturalistic mode is a set-up: the conversation is staged to *appear* natural. In the intense dialogue with machines, there is no room for the two human actors' emotional improvisation or personal response. As pointed out by theater scholar Francesca Spedalieri in a review of a Hirata production (2014), every single word, every slight movement, is conceived, staged, directed, and rehearsed. In Hirata's play, the actors do not need to have a heart, they just need to follow the script. This approach seems to correspond exactly with Ishiguro's views as an engineer when he creates robots, as discussed above. The audience may reflects upon to which degree the human actors actually differs from the robotic actors when they realize that the human actors are programmed and controlled in a way similar to the robots. Naturalism as an aesthetic framework becomes an experimental form; it is used as an epistemological tool for reflections that makes us aware of broader issues, such as questioning the fundamental issue of what is "natural" or "normal" in human beings.

Differentiation of Citationality

This points back to Austin's phrase about the "parasitic use" of language and the critique this particular part of his theory has generated. In his text "Signature Event Context", the philosopher Jacques Derrida (1988) offers a productive elaboration on Austin's remark. Rather than dismissing the infelicities uttered by an actor on stage or introduced in a poem such as Austin seems to do, Derrida turns the argument around and instead suggests that there is a citational element, an iteration, in all performatives. The words "I do" in the wedding ceremony is felicitous because the utterance is coded through repetition: the words have been uttered before, within an iterable set of circumstances and context. All performative speech is in some sense always a citation. Derrida does not dismiss citationality as a criteria for infelicities, instead he proposes a differentiation of citationality. He speaks of a "relative purity" of performatives that "does not emerge *in opposition* to citationality or iterability, but in opposition to other kinds of iteration within a general iterability" (Derrida 1988, p. 18). In similar veins, the robots in *I, Worker* appear at first sight to be an infelicitous and parasitic use of the norms of the human. However, using Derrida's argument, it is possible to understand the robots as iterations or citations of human beings in movement, behavior, and speech. The performative appearance of the robot is not in opposition to the "purity" of human intention and emotion, but in opposition to other kinds of iteration within a general iterability. If robots are citations of humans, humans too are citations of other humans; of iterable social norms and conventions.

This aligns with theories of performativity proposed by philosopher and gender theorist Judith Butler in her book *Bodies That Matter* (1993), in which she argues that the subject is constituted by performative repetitions of social codes. The iterative practices that make up the performative elements in social interaction constitute the subject with very little opportunity to act differently. Butler uses the social construction of gender as her basic argument, and points out that the process gendering a body within the sex/gender system is performed by social norms outside the subject, and is therefore a performative process outside the control of the individual. Formation of a subject happens because the individual repeats acts of social norms, understood as a process of citations of previous acts. This, however, is not to say that every body is submitted to the processes of "becoming" (for example being gendered) entirely without agency. In the iterative processes, there is an instability in the repetition because two acts can never be the same, and this aspect of instability provides potential for agency and power. Butler speaks of "sedimented effect of a reiterative or ritual practice" that has a naturalizing effects, but it is at the same time by virtue of this reiterative practice that agency is possible because "gaps and fissures are opened up as the constitutive instabilities in such constructions, as that which escapes or exceeds the norm, as that which cannot be wholly defined or fixed by the repetitive labor of that norm." (ibid., p. 10). Butler points out the possibility for change in between each repetitive act as the power that undoes the stabilizing effect. The slight displacement of the previous act on to the next allows for escape or excess of the norms.

The Glitch

In an ironic way, this is exactly what happens in the DVD version of a recording of *I, Worker* (Robot Theatre 2010). In the recorded scene described above where Yūji's apology to the male robot Takeo leads the other robot to state that "robots cannot lie", there is a slight displacement. After the moment where Yūji has apologized and the male robot proclaims that it not Yūji's fault, the robot adds "hontō desu" ("it is true"). However, this is where a technical glitch happens: the sentence to follow, "robotto wa uso wa tsukemasen kara" ("because robots cannot lie"), is spoken by the female robot exactly at the same time. Something in the timing of the two robots' speech acts has gone wrong, the result of which is that the two robots appear to be speaking all at once, and both utterances thus become unintelligible. This is a somewhat awkward moment in the play, which is otherwise characterized by silence and long pauses in the dialogue. It is not noted anywhere whether the robot voices are in fact human voices mediated live from behind the scene (as Wizards), or if the robot voices are pre-recorded and played back during the performance. Perhaps it was a case of human error in the programming, and the glitch revealed how difficult it is to construct "natural" conversations between human and robots. Whatever the case, the consequence of the slight displacement in the DVD version is that the two robots appear to be competing for the right to speak; they do not seem to have the appropriate social skill or polite attitude to let the other speak first. The displacement reveals an agency on part of the robots because they do something unexpected (something that is not in the script), and they escape and exceed sociocultural norms (such as showing politeness by letting others speak first). It makes the viewer aware of the naturalized effect of the conventions of politeness in conversation, the rules of not interrupting when the other speaks. The glitch becomes one of the "gaps and fissures" suggested by Butler: it is what induces productive instabilities of human rituals and un-does stabilizing effects.

So let me finish by returning to the ambiguous and paradoxical statement uttered by a robot on stage: "robots cannot lie." Of course robots lie. They do so every time aspects of human behavior, social intelligence or emotional response are attributed to the machine, whether in a robot–human theater production, or in simulated robotic home companion tests. The robot acts as a "parasite" on human interaction, because the robot is attributed intention by the human. Such performance of human-ness by something other than human is "in a peculiar way hollow or void", as Austin would have said, but these ways "parasitic upon its normal use" is exactly what art contributes to the field of robotics: it opens for new insight of how humans project and respond to their own behavior and emotions in their interaction with technology. Human beings, like robots, are constituted through a kind of social "programming" that makes humans behave within certain norms and conventions. Theories of performativity and aesthetics show how such conventions may be negotiated and provide agency for robots and humans alike. New forms of interaction may emerge.

References

Austin, J.L. 1975. *How to Do Things with Words*. Cambridge, MA: Harvard University Press.

Barnes, J.A. 1994. *A Pack of Lies: Towards a Sociology of Lying*. Cambridge: Cambridge University Press.

Borggreen, G. 2014. ""Robots Cannot Lie": Performative Parasites of Robot–Human Theatre." In *Sociable Robots and the Future of Social Relations: Proceedings of Robo-Philosophy 2014*, edited by Johanna Seibt, Raul Hakli and Marco Nørskov, 157-63. Amsterdam: IOS Press Ebooks. doi: http://dx.doi.org/10.3233/978-1-61499-480-0-157.

Breazeal, C., A. Brooks, J. Gray, M. Hancher, J. McBean, D. Stiehl, and J. Strickon. 2003. "Interactive Robot Theatre." *Communications of the ACM - A Game Experience in Every Application* 46(7): 76-85. doi: http://dx.doi.org/10.1145/792704.792733.

Butler, J. 1993. *Bodies That Matter: On the Discursive Limits of "Sex"*. New York: Routledge.

Carlson, M. 1996. *Performance: A Critical Introduction*. New York: Routledge.

Chatley, A.R., K. Dautenhahn, M.L. Walters, D.S. Syrdal, and B. Christianson. 2010. "Theatre as a Discussion Tool in Human–Robot Interaction Experiments: A Pilot Study." *Third International Conference on Advances in Computer–Human Interactions, 2010*. ACHI '10 10-15 Feb. 2010. 73-8. doi: http://dx.doi.org/10.1109/ACHI.2010.17.

Derrida, J. 1988. "Signature Event Context." In *Limited Inc.*, edited by Gerald Graff, 1-23. Evanston: Northwestern University Press.

Eckersall, P. 2011. "Hirata Oriza's Tokyo Notes in Melbourne: Conflicting Expectations for Theatrical Naturalism." In *Outside Asia: Japanese and Australian Identities and Encounters in Flux*, edited by Stephen Alomes, Peter Eckersall, Ross Mouer and Alison Tokita, 237-43. Melbourne, Vic.: Japanese Studies Centre. Accessed 2014/05/22. http://hdl.handle.net/11343/32503.

Hirata, O. 2010. "Hataraku Watashi" [I Who Work]. In *Robotto engeki [Robot theatre]*. ed. Center for the Study of Communication-Design Osaka: Osaka daigaku shuppankai.

Hirata, O. and M.C. Poulton. 2002. "Tokyo Notes: A Play by Hirata Oriza." *Asian Theatre Journal* 19(1): 1-120.

Inobeeshon 25. 2008. "Yume No Aru Mirai No Jitsugen No Tame." [Innovation 25. For the realization of the future of our dreams]. Cabinet Office, Government of Japan Accessed February 3, 2015. http://www.cao.go.jp/innovation/action/conference/minutes/20case.html.

Ishiguro, H. 2010. "Robotto Kenkyū Towa Ningen No Kokoro O Shirukoto" [to Know the Human Heart through Robot Science]. In *Robotto engeki [Robot theatre]*. ed. Center for the Study of Communication-Design. Osaka: Osaka daigaku shuppankai.

Jakovljevic, B. 2002. "Shatterede Back Wall: Performative Utterance of a Doll's House." *Theatre Journal* 54(3): 431-48.

Lu, D.V. and W.D. Smart. 2011. "Human–Robot Interactions as Theatre." RO-MAN, 2011 IEEE, Atlanta, GA, July 31 2011-Aug. 3 2011. 473-8. doi: http://dx.doi.org/10.1109/ROMAN.2011.6005241.

Riek, L.D. 2012. "Wizard of Oz Studies in HRI: A Systematic Review and New Reporting Guidelines." *Journal of Human–Robot Interaction* 1(1): 119-36. doi: http://dx.doi.org/10.5898/JHRI.1.1.Riek.

Robertson, J. 2007. "Robo Sapiens Japanicus: Humanoid Robots and the Posthuman Family." *Critical Asian Studies* 39(3): 369-98. doi: http://dx.doi.org/10.1080/14672710701527378.

Robertson, J. 2010. "Gendering Humanoid Robots: Robo-Sexism in Japan." *Body & Society* 16(2): 1-36. doi: http://dx.doi.org/10.1177/1357034X10364767.

Robertson, J. 2014. "Human Rights Vs. Robot Rights: Forecasts from Japan." *Critical Asian Studies* 46(4): 571-98. doi: http://dx.doi.org/10.1080/14672715.2014.960707.

Robot Theatre. 2010. *I, Worker*. Tokyo: National Museum of Emerging Science and Technology.

Schechner, R. 2006. *Performance Studies: An Introduction*. New York: Routledge.

Spedalieri, F. 2014. "Quietly Posthuman: Oriza Hirata's Robot-Theatre." *Performance Research* 19(2): 138-40. doi: http://dx.doi.org/10.1080/13528165.2014.928530.

Syrdal, D.S., K. Dautenhahn, M.L. Walters, K.L. Koay, and N.R. Otero. 2011. "The Theatre Methodology for Facilitating Discussion in Human–Robot

Interaction on Information Disclosure in a Home Environment." RO-MAN, 2011 IEEE, July 31 2011-Aug. 3 2011. 479-84. doi: http://dx.doi.org/10.1109/ROMAN.2011.6005247.

TPAM Direction. 2011. "TPAM Direction: Masashi Nomura Program." *TPAM in Yokohama: Performing Arts Meeting in Yokohama 16th (Wed) - 20th (SUN), February 2011*, program sheet, Yokohama: TPAM Yokohama.

PART III
Challenges

Chapter 9

Robots, Humans, and the Borders of the Social World

Hironori Matsuzaki[1]

This chapter investigates fundamental border issues of the social world, which are triggered by the development and diffusion of autonomous human-like robots. From a sociological standpoint, I discuss the question of how anthropomorphic robots will challenge the modern concept of social actors and have crucial impacts on current institutional orders of human society. In a further step, I outline an analytical framework for empirical analysis of such elementary border phenomena, by referring to the core aspects of sociality shared in the field of social theory. After discussing a methodological problem of dyadic approach, I claim that the basic model of social interaction should be explained from a triangulated interaction between three embodied selves.

Introduction

In modern democratic societies for which the ethos of human rights is of fundamental importance, it is taken for granted that only living human beings are legitimate members of society, i.e., *persons*. Every human being is supposed to be treated both ethically and legally as such, even if she[2] is incapacitated from performing conscious actions (e.g. through a serious brain damage). Those who are in coma or suffering from severe dementia can still count on treatment as individual persons, as long as they are diagnosed as being alive. Being a living human is identified as being social, which in turn means that all other entities are deemed out of this range. Although nonhuman entities are integrated in many areas of life, there are currently borders that they cannot cross in a generally accepted way.

1   This chapter presents results from the research project "Development of Humanoid and Service Robots: An International Comparative Research Project—Europe and Japan", funded by the DFG German Research Foundation. I would like to thank the DFG for their support.
2   In this chapter I alternately use masculine and feminine forms for the singular of the second person in order to maintain gender neutrality.

In recent years, however, the equation of the realm of the social with the world of humans has been increasingly challenged by the rise of robots that resemble humans in physical appearance and/or perform tasks in much the same way as humans do. Many of them are developed to solve practical problems of daily life in place of humans and should become—at least some extent—autonomous. They are intended to be independent entities that controls themselves and improve their own behaviors based on "experiences" (Matthias 2004). This is regarded as a necessary condition in order for them to undertake functional roles, which only humans could afford to play. Accordingly, there is a growing concern that practical applications of and interactions with these robots could pose a variety of challenges to society (e.g. Foerst 2009, Darling 2012, Coeckelbergh 2010). Should autonomous human-like robots be seen as entities that do more than just simulate functional roles of human actors? If so, does their presence in everyday life affect the concepts of agency, personhood or companionship? How should they be treated? Are they agents that can be held accountable for what they have caused?

These developments draw attention to the contingent nature of sociality. Findings in ethnological and historical research (Kelsen 1943, Luckmann 1970) have revealed that the anthropocentric notion of personhood in Western modernity does not hold true throughout the ages and cultures. That is, the premise that only living humans can be legitimate social actors cannot have permanent validity; the sphere of persons would rather be demarcated as a result of historically and culturally contingent processes of interpretation. If one follow this view, it is likely that everyday interactions with robots with features nearing to those of humans will urge reconsidering the concept of what it takes to be a social actor. Surprisingly enough, however, the question of sociality is not sufficiently addressed in HRI and social robotics. Rather than examining sociality in its fundamental definition,[3] many studies in this research area start with an implicit assumption: when interacting with ordinary people, robots should be seen as part of 'social' interaction—simply on the grounds that humans are involved (see e.g. Alač, Movellan, and Tanaka 2011). Given the range of impacts these robots may have onto society, a more systematic approach is needed.

The question I am mainly concerned with in this context is how to conduct a critical (field) research on border problems posed by anthropomorphic robots. Drawing on Lindemann (2005, 2009a), I take here an alternative approach. As a first step, one must desist from postulating that only humans can be social actors, and at the same time, keep reflexively a distance from anthropological assumptions concerning human/nonhuman distinction. A human–robot interaction should be analyzed in the same way as human–human interactions. However, doing so poses a crucial methodological problem: how can we identify the preconditions of social interaction without recourse to anthropological knowledge and perspectives? I suppose this problem can be solved by elaborating a formal theory of the social. In sociology, there is an implicit consensus on the basic properties of what is to be

---

3   For a few exceptions, see Pfadenhauer (2014) and Meister (2014).

seen as social phenomena. This abstract notion can be used to develop a general concept of the social, through which practical interactions with robots as well as their agency can be empirically analyzed.

The chapter is organized as follows. First, I briefly review the interrelation between functional differentiation of modern society and the border regime, according to which the realm of the social is delimited. This will be followed by a discussion of how autonomous robots with anthropomorphic features will challenge the fundamental institutions of modern society by causing irritations to the existing legal norms. I then try to expound the analytical framework for empirical research on such border issues. This will be unfolded in the following steps: (1) I begin with a consideration of what should be regarded as constitutive for sociality. The key idea consists in the focus on the problem of double contingency, which emerges within the interaction between the "embodied selves" (Ego/Alter). The solution of this problem entails the generation of a reified order that guide their relationship and can be seen as an essentially social phenomenon. (2) In a second step, a methodological problem will be discussed that will occur if the basic model of sociality is founded on a dyadic constellation. Social interaction is driven by reciprocal understanding of the other, which presupposes an initial decision on both sides: participants mutually recognize the other as a you who will interpret me. If a questionable entity appears in interaction, it must be proved whether this entity can be trusted as an actual "you." However, this question can only be answered with reference to the perspective of a third party (Tertius). (3) With this in mind, I finally claim that the core features of the social should be explained from a triangulated interaction between three entities (Ego/Alter/Tertius).

Robots and Problems of Social Order

Since its beginnings, sociological research has been motivated by the question of how social order is possible (Simmel 1911, Luhmann 1985). In everyday life, an order which forms society is more or less taken for granted. For sociologists, it is something that requires explanation. The problem of social order necessarily contains the problem of borders of the social world. A society always requires a shared concept of who is entitled to participate in the generation and reproduction process of its institutions (and who is not). The question is not how particular entities are integrated in the areas of life, but rather whether or not they can be recognized as members acting as subjects in the formation process of social order. Such understanding becomes a significant part of reality for those who reside in the realm of cohabitation. This notion is derived from the theorem that societies can be differentiated according to the features of how the sphere of social actors is delimited (Lindemann 2009a). Different societies develop different concepts of which entities fall into the classification of social actors and which not. For a sound understanding of institutionalization process, it has to be taken into account where and how this type of inclusion/exclusion principle is constructed.

In most present-day societies, this question is answered based on a generalized criterion. There is a broad consensus that only living human beings can be legitimate members of society. They are thought to deserve this status, simply by virtue of being human. For nonhumans, by contrast, there are currently borders they cannot cross in a generally recognized way. For instance, machines are not accepted as subjects of legal rights and duties. A computer is excluded from the category of responsible actors, although some lay user occasionally anthropomorphizes its features and assigns them greater agency than it might have (Nass and Moon 2000).[4] The same is true for other-worldly beings such as angels, devils, and spirits. As a result of the disenchantment of the world (Weber 1946) which occurred in association with secular rationalism and advances in natural sciences, constant relationships with other-worldly beings (e.g. a deal with the devil) became irrelevant to society at large (Neumann 2007).

These facts represent core aspects of Western-inflected modernity, which are in large part institutionalized and effective among social actors at multiple levels. It is nowadays a dominant reality of everyday life that only living humans recognize each other as those who they should make sense of and respond to. The normative fundament behind it is the idea of universal human rights, which is enshrined in the milestone declarations as represented by the Universal Declaration of Human Rights (United Nations 1948). These foundational documents assert that all members of the human family are entitled to human rights and basic freedoms. They are guaranteed the inherent dignity as well as the right to be treated equally, without distinction as to ethnic and cultural backgrounds or religious belief, irrespective of sex, age or language. The formal equality before the law is distilled to the status "person" (ibid., art. 6). Here, a normative framework is provided through which universal protection of human rights (or, to be more precise: fundamental rights exclusively granted to human beings) is promoted. The identification of all human individuals with legitimate members of "our world" rests on an institutional complex resulting from the ethos of human rights.

This order is inextricably intertwined with another characteristic feature of modern society: functional differentiation (Luhmann 1965, see also Verschraegen 2002). The structure of modern society is differentiated according to significant functions of social life and subdivided into specialized domains such as economy, science, law, art, religion, politics, education, medicine, etc. The question inevitably arises here as to which entities are permitted to join activities within these functionally specific domains. In terms of mobilization of elements of the social, the universal validity of human rights plays a significant role. The underlying idea is that all living humans should be universally recognized as those who are eligible for membership of the sphere of persons. This focus in turn means that only these entities become driving forces of practical order formation, which

---

4   This becomes visible particularly in that technological artifacts are—at least at present—not counted as part of the category of those who are eligible for compensation for loss or damage (see below).

can be mobilized in the same way for any given interaction. That is, functional differentiation of society is founded on the premise that the sphere of social actors is staffed by individuals of the same kind—namely: congeneric human beings. Anyone who is recognized as a member of the human species can have, as a general rule, an equal access to all these functional domains of society and receive treatment as an actor in its own right. They are considered agents who are responsible qua individuals, capable of making and acting upon rational decisions, and interact with other such actors in their own networks. Seen in this light, it is reasonable to say that the current understanding of the term "human being" is in itself an institution which both normatively and cognitively, serves as a base for the functional differentiation of modern society.

Decisive in terms of status of nonhumans (including robots) is a flip side of the equation "all humans are persons." Institutions deriving from the ethos of human rights are based on a clear distinction between human actors and those entities, which are deemed not to bear the criteria for being a biologically living human being. This distinction is crucial for relationships in the ordinary world because it affects an overall decision as to which types of treatments are appropriate for particular entities. All nonhumans are ruled unfit to be driving elements of the social and thus excluded from the reproduction process of institutional orders. For example, it is basically not assumed that a computational artifact qualifies as a lawyer or a candidate for political election, regardless of the level of autonomy and intelligence it may have; under current laws, a pet animal is not regarded as being eligible for social security even if it may clam the right to receive humane treatments for its well-being. In this respect, the categorical demarcation between humans and nonhumans provides a primary criterion for sociation, continuous maintenance of which is of fundamental importance for the formal structure of modern society.

With a focus on the relevance of human/nonhuman distinctions for social differentiation, Lindemann (2009a, b) proposes the concept of the "societal border regime." She argues that the fundamental structure of society can take on different forms depending on how the sphere of persons is delimited. In Western modernity, as seen above, the "this-worldly living human" serves as a point of reference on this matter. This gives rise to the question of how the understanding of being a human is constructed. According to Lindemann, modern society is built on a four-way demarcation process, thorough which human beings are distinguished from other entities: the two borders which should mark the onset and the end of a human life, and the two borders that separate humans from machines and other animals. The former two borders are constructed as ones that can potentially be transgressed, and thus lead to well-known boundary questions posed in the contexts of abortion and brain death. At what point is a human being alive enough to deserve the right to life? At what point is a human being no longer alive enough to receive treatment as a person? By contrast, the latter two boundary definitions concern borders that basically cannot be crossed: the human/machine distinction and the human/animal distinction.

As a consequence, the human being is understood "as a living, earthly being who begins to live at an identifiable point in time for a limited duration, is not on the same level as an animal, and is not a machine" (Lindemann 2010, p. 285). This understanding is the cognitive condition for being able to refer to the "human" in a general sense and to communicate with each other as part of the universal community of persons. As such, human individuals treat each other and act in line with the logics and requirements of particular functional domains. The fourfold differentiation, which Lindemann refers to as "anthropological square", makes it possible to establish binding criteria for membership of the sphere of social actors. Thus, the biologically living human being can be seen as the core institution of modernity, which defines and delimits the realm of the social by demarcating human individuals from other entities.

For the modern border regime, it is crucial to maintain strict distinctions between humans and machines (or animals) by means of institutional criteria. This does not exclude the possibility that nonhuman entities may have significant effects on inter-human relationships. In fact, they are nowadays constantly mediated by a technological infrastructure which can impact significantly upon them. Material effects of technology are, however, perceived as not having a social character; that is, technological artifacts exert their effects in a different way than human actors do in relation to one another (Lindemann 2011). Nevertheless, it is important to recognize that the anthropological square does not represent permanently-fixed boundaries, but rather, social institutions, which have been developed in historically contingent ways and thus can potentially be contested (Lindemann 2005, 2009b).

The initial equation of the realm of the social with the world of humans has been subjected to study in sociology from early on (see Kelsen 1943, Luckmann 1970). As empirical findings in cultural anthropology reveal, premodern societies regarded not only human beings as legitimate social actors, but also plants, animals, deities, and/or deceased humans. A historical example to mention in this context would be animal trials in medieval Europe, in which nonhuman animals, including insects, were involved in criminal proceedings (Evans 1987). Animal defendants were arraigned before church or secular courts on suspicion of having committed a crime (e.g. homicide), and if convicted, executed or exiled from the region concerned. The criminal trials against animals did not differ from those against human defendants in how the defendants were treated in court proceedings. The accused animals were considered subjects endowed with moral agency and thus could be held culpable for a criminal act in the same way as humans. Such trials remained part of several legal systems until the early 18th century. The features of a social counterpart, in premodern perceptions still attributed to nonhumans, have lost its efficacy in the course of modernization. Relationships with nonhumans have, over time, been desocialized.

Similarly, the problem of social actor has been discussed by approaches that theorize the agency of technological artifacts, such as Actor-Network Theory (Latour 2005, Callon 1986, Cerulo 2009) or the theory of distributed agency

(Rammert and Schulz-Schaeffer 2002, Rammert 2012). These studies raise the question of whether nonhuman entities, especially technological artifacts, should also be taken into account as actors in terms of material effects for the stability of social orders. Based on a criticism of the marginalization of nonhuman agency in social theory, they question the validity of the notion of human beings as the only possible form of social actor. Technological artifacts should be—at least to some extent—assigned the status of a sociologically relevant actor, considering that it can contribute to the accomplishment of (collective) action by producing certain effects. However, this claim fails to delve into the heart of the border issue in that it only indirectly touches on the question of how the sphere of legitimate social actors is delimited. According to the theoretical models proposed here, nonhumans 'act' only in a less rigorous sense. At the same time, it is implicitly assumed that human beings should be given a privileged position: technology may have the function of guaranteeing the stability of social orders, but sociality itself is determined by reference to human agency; as such, humans can attribute (or deny) other beings the status of social actor (Lindemann 2008, Matsuzaki 2011). This indicates that these approaches are bound by the perspectives of modern society according to which only living human beings can be persons.

An alternative would be to take seriously historical forms of sociality in nonmodern societies, which differ from those in Western modernity. The understanding of who (or what) can be social actors varies across the ages and cultures. In other words, the borders of the social world are determined based on the historically and culturally contingent process of interpretation. If one follows this argument, it opens up a new opportunity for considering possible impacts of human-like robots onto the modern border regime.

In recent years, the exclusive status of humans has been increasingly questioned by the development of robots that physically resemble humans and perform activities as humans do. These technological artifacts challenge the modern concept of personhood in that they blur the distinctions between humans and machines in many respects. This can be well-observed in humanoid robotics where the artificial reconstruction of human features is the core objective of research. Here, the human body serves as a source of inspiration for the design of anthropomorphic robots or hybrid-bionic systems (neuroprostheses, robotic exoskeletons, etc.). This enterprise is primarily motivated by the desire to explore the conditio humana through interdisciplinary collaborations (e.g. Ishiguro and Nishio 2007, Jank 2014, MacDorman and Ishiguro 2006). Using the developed platform, researchers in robotics and other disciplines try to validate scientific theories about the functions and physical characteristics of the human body or investigate experimentally socio-psychological effects of inter-human interaction. The impact of humanoid robotics goes beyond the framework of the "engineering anthropology." It is generally expected that in an aging society, robots will increasingly be used in everyday life situations as service providers that can operate independently (Feil-Seifer, Skinner, and Matarić 2007). Some of them have humanlike features and simulate emotional expressions and/or other aspects of inter-human communication, in order to facilitate the smooth interaction with lay users (Zhao 2006, Yamazaki et al. 2012).

There also seems to be good reason for reflecting on extreme frontiers of technological advance, as represented by biologically-inspired and/ or neuroscientifically-informed robotics. Biologically inspired robotics has become an active area of research, mainly because of the assumption that the electromechanical simulation of biologically living organisms (animals, insects or plants) will have potential to provide a major contribution to the solution of many technological problems (Menciassi and Laschi 2012). Moreover, there is a strong interest in implementing neurobiological expertise into automation technology, which is partially reflected in the conception of personal service robots. Some personal care/companion robots are designed as entities that would be endowed with a certain level of sentience and thus could respond to various needs of vulnerable people in the same way as care personnel do.[5]

These developments raise the question of which status is appropriate for robots with bio- or anthropomorphic features. It is assumed that applications of such artifacts in everyday environments would cause confusion concerning social institutions, particularly due to the features which distinguish the robot from other machines: its role as an interaction partner, a high degree of autonomy, and ability of self-directed learning. The robot increasingly moves beyond the reach of control and oversight by engineers, as its independence in operation grows and behavioral patterns of the system constantly change (Matthias 2004). The classification of the robot as a mere object turns out to be inadequate, and it becomes difficult to directly apply currently existing institutional norms and principles. Such confusion would be all the more problematic when it comes to the question of robot's general status. Should an autonomous robot be considered an entity that can be held responsible for what it has caused? Is it reasonable to assume that robots would be generally recognized as persons? Can they enjoy protection and guarantees similar to those provided for humans? Last but not least, this irritation leads to a more fundamental question: what does it mean to be social?

The robot with the label "more than a mere machine" is, in fact, increasingly discussed in the field of legal sciences as well as in the debate on the extension of moral agency to nonhumans. Following the findings in science and technology studies, it is often argued that a strict functional differentiation between human

---

5   A typical example can be found in Robot Companions for Citizens, one of the candidate projects for the Future and Emerging Technologies Flagships funded by the European Commission. See http://www.robotcompanions.eu (accessed on April 29, 2015). The project aims to contribute to the creation of a sustainable welfare through the development of sentient robots that integrate "perception, cognition, emotion, and action with a contextual awareness of self, others, and the environment." These machines are developed to become independent entities that not only simulate external appearances and mimic behaviors of humans, but also incorporate internal features, which are considered a prerequisite for being a reliable partner in interaction.

beings and technological artifacts ("man-machine dualism") is no longer acceptable (e.g. Adam 2005, Verbeek 2006, 2009). Instead, contributions of both humans and things (including machines) are seen as inextricably linked together in a hybrid network from which actions and decisions arise. In those approaches, mental states (intentionality, free will, emotions or the like) are not deemed necessary requirements for an entity to qualify as an agent; but rather, the focus is on what follows from the assemblages of humans and nonhumans (Akrich 1992, Latour 1992). Since tasks are distributed among multiple entities that mutually affect each other in contingent ways, it becomes difficult to retrospectively identify who or what is responsible.

Against this background, Floridi and Sanders (2004) propose to extend the class of moral agents to include complex AI systems, while disconnecting agency and accountability from the notion of responsibility. In their view, computational artifacts can be the cause of a morally charged action and thus should be acknowledged as moral agents, which can be held accountable, but not responsible. Johnson (2006) rejects the idea of ascribing independent moral agency to computer systems. She claims instead that computational artifacts can be relevant entities for normative evaluation only in that designers and developers inscribe in these artifacts their particular values and intentions, which will be activated by the interaction with human users. Researchers within the field of machine ethics develop discussions on whether and how a given ethical framework, such as Kantian or utilitarian ethics, can be implemented into machines (Allen, Smit, and Wallach 2005, Powers 2011, Tonkens 2009, Torrance 2008, Wallach and Allen 2009). On the basis of such a normative model, according to the authors, computational machines like robots could become surrogate agents that 'make ethically relevant decisions' on behalf of its developers or users.

The effort to apply normative constraints to machines raises the question of whether complex computer systems should be given a particular legal status, which would allow for holding them responsible and liable. Some authors insist that it makes no sense to treat computer systems (e.g. robots) as entities that can be held legally responsible, for that they do not suffer and thus cannot be punished (Sparrow 2007, Asaro 2012). Others argue that there is, at least theoretically, no fundamental reason why nonhuman entities, including robots and other computational systems, could not be viewed as subjects of legal rights and/or duties (Teubner 2006, Calverley 2008). The possibility of attribution of legal personality to contracting software programs (under a civil law framework) has been discussed for many years (e.g. Solum 1992, Karnow 1994, Allen and Widdison 1996, Andrade et al. 2007). In the countries featured, there are no special regulations that would explicitly grant legal or contractual capacity to a software program. To recognize legal personality of computational entities requires either expanding the current legal framework or creating a completely new one. For this reason, the discussions mainly focus on the question of how the existing legal institutions, e.g. the rules on "messenger", "minor" or "representative" (in terms of agency law), could be analogously applied to software programs in electronic transactions. The result is that electronic entities could have, if any, a minimal status under the existing legal norms; responsibility and liability fall back after all on natural persons or corporate bodies. As an alternative approach, Wettig and Zehendner (2004) advocates the concept of "electronic person" in terms of a legal entity with limited liability, and propose to establish an "agent liability funds" assets of which could be issued to those involved in case of problems caused by the use of automated machines (see Stahl 2006, Matthias 2008, Beck 2013 for similar concepts).

Two Different Interaction Frames

As seen above, the critique of the modern border regime shows that the understanding of "biologically living humans" as the only vehicles for sociation process is to be regarded as a historically contingent phenomenon specific to modern society. This finding on the level of theory of society suggests that being social cannot be identified a priori with being human. On the level of social theory, the contingency of the social world must be taken seriously for the analysis of empirical material. The questions arise here as to how the delimitation of the social concerning anthropomorphic robots will appear in the real world, and how such border phenomena can empirically be studied without falling into the pitfalls of anthropological biases. A research on this subject requires, both normatively and cognitively, a critical distance from those perspectives, which are products of the modern border regime. In terms of methodology, this implies a crucial problem. Human–robot interactions should be analyzed in the same way as interactions between humans. In order to meet this requirement, a sociological observer needs to be free from a priori assumptions, which are based on anthropological knowledge. The preconditions of social interaction have to be redefined without recourse to human/nonhuman distinctions.

I suppose that this can be achieved by elaborating a formal theory of the social. In sociology there is an implicit consensus on the basic properties of what is to be seen as social phenomena (Lindemann 2005). In spite of conceptual diversities, many social theories converge on this point when it comes to the core aspects of sociality. This abstract notion can be used to develop a general concept of the social, through which practical interactions with robots as well as their agency can be empirically analyzed. In what follows, I will outline the framework of this approach.

Ego-Alter Structure (Dyad): Simple Interaction Frame

I begin with a consideration of what should be regarded as constitutive for sociality. The key idea consists in the focus on specific complexity which emerges within the relationship practically conducted between the two embodied selves (Ego/ Alter). A self is here conceived of as an entity that has a body (1); perceives events in its environment as well as its internal states (2); takes an action on practical requirements in a particular situation (3); is aware that it mediates perceptions and actions by itself (4); and finally can develop expectations on what will happen in the environment as well as on what the others are expecting (5).

Ego, an embodied self, experiences that it is in relationship with Alter, who is another embodied self. Both participants assume that bodily events of the other side (e.g. Alter for Ego) are dependent on the way in which this entity mediates between perception and action. They relate to each other by expecting and interpreting the expectations/intentions of the respective counterpart. To put it precisely: Ego incorporates the behavior of Alter into its own behavior, by anticipating that Alter expects Ego to make its own behavior dependent on Alter—and vice versa. If both sides try to accord their own actions with the expected expectations/intentions of the other side, their behaviors become conditional on each other in a highly complex way. A twofold uncertainty emerges: neither Alter nor Ego can be certain what to expect in the next instant. This uncertainty, called "double contingency", means a practical problem to both participants, which must be solved within their interaction (Parsons 1968, Luhmann 1995).

The solution of "double contingency" entails the generation of a specific order, which guides the participants on how to interpret expectations and perspectives of the respective other side. In a first step, one participant would take a tentative action meaning something, e.g. with a glance or a gesture. Subsequent steps referring to this first step would lead to activities, which serve to reduce contingency in interaction. If this works out, both participants would be able to build up concrete expectations about the future course of their mutual activities. As a result, a system of symbolic meaning emerges, which can be drawn on by both sides. Starting from this point, further mechanisms could be developed to generate an order that not only allows the participants take actions and shape perspectives of the world surrounding them, but also mediates their relationship appearing in various ways such as confidence, rules, obligations, and symbols. This mediating order is to be understood as an essentially social phenomenon because, once it has emerged, it cannot be reduced to the activities of a single party (Simmel 1992, Weber 1980, Schütz 1974, Berger and Luckmann 1991). It manifests itself as a factual structure that can be shared by both participants.

A Methodological Problem of the Dyadic Model

If one follows this conception, a mediating social order can emerge from the complex interdependence between embodied selves, which appears, at its most basic level, in the situation of double contingency. However, a methodological problem occurs if the model of sociality is founded on a dyadic interaction. Take, for instance, a bipartite agreement that becomes factual between two actors. Its efficacy outside the dyadic framework of relationship is basically not guaranteed. More importantly, a mutual arrangement within a dyadic relationship is unstable by nature. One participant (Ego) can take back or change his promises at his own discretion, even if the other (Alter) claims that these can be terminated only when agreed by both sides. Such a problem is probable, especially when it comes to a conflict between the two parties. Therefore, mediating orders within a dyadic constellation need to be stabilized with reference to the perspective of a third party (Luhmann 1985). By the same token, the dyadic Ego-Alter relationship itself needs to be objectified in order to become a reality beyond its own frame.

The Ego-Alter interaction is driven by their reciprocal critique of one another's physical appearance and symbolic utterance, because there is basically no immediate access to the inner state of the other. This implies the distinction between expressive surface and non-appearing inside (Plessner 1975). Expressive surface of an embodied entity indicates that there is a non-appearing inside of this entity. By this I mean those phenomena which reside within a body, but cannot themselves be materialized—such as self-consciousness, consciousness or state of being alive. The directly observable outside and the non-appearing inside are inextricably linked with each other, while the latter becomes recognizable only with reference to the former. That is, core properties of an entity (e.g. "alive") are expressively realized in the form of bodily events it accomplishes. The assessment of the latent side is decisive for interaction in that it affects the decision on how to deal with a particular entity; this can be made only through an observation of expressive surfaces. Social interaction always includes the interpretation of other's non-appearing inside.

In this context, it is necessary to logically distinguish two steps of interpretation. The interpretive procedure between two actors presupposes an initial decision on both sides. Ego and Alter do not only interpret expectations, but mutually recognize the other side as a *you* who will interpret *me* (Lindemann 2005). That is, they make—in the first place—a judgment as to whether their respective counterpart is one whose expectations are to be expected *or* an entity which do not need to be treated in such a manner. Only if they accept each other as another self, the process of reciprocal understanding can take place. This double interpretation is a fundamental characteristic of the interaction, which can be understood as *social*. In case a questionable entity appears in Alter-position, it must be proved whether or not this entity can be trusted as a reliable counterpart who can participate in a You-and-I relationship.

Within a dyadic interaction, however, idiosyncratic personification is possible in principle (ibid.). One can arbitrarily treat an object as a you by interpreting it as being endowed with the inside that is to be understood (e.g. mental states), although this interpretation is not true when viewed in light of the generally accepted criteria. He may even try to develop an intimate relationship with this entity. Such a peculiar interpretation is not uncommon in everyday human-to-thing interactions. A number of quasi You-and-I relationships can be created: a bond between a fish and me, a bond between a mannequin and me, etc. But usually these types of relationships only rest on individual views from the Ego-position, which cannot verify the authenticity of another self in a general sense. Thus, it remains unclear whether Ego's counterpart is an *actual you* or something else. After all, this question cannot conclusively be answered within the dyadic structure of interaction. For a coherent theoretical explanation, it requires a third figure: Tertius.

Ego-Alter-Tertius (Triad): Objectified Interaction Frame

With this in mind, I claim that the core features of the social should be explained from a triangulated interaction between three selves. In order to become a stable dyad, the relationship between Ego and an entity X needs to be exposed to the perspective of a third party. The inclusion of a third party has two significant implications:

(1) The problem of egocentric interpretation will be resolved if the Ego-X relationship is under observation by an external party. If the presence of another self becomes an issue, this needs to be examined with reference to the perspective of a third actor. For this to occur in practice, an "initial spark" must be given in the first place. Someone in Ego-position (actor A) perceives particular features X shows in the interaction and interprets them as indicating that this entity may be seen as a potential Alter. This actor suggests the idea to someone else (actor B) and involves him/her into the situation. Both actors (A & B) then communicate with each other as to how X relates itself to other entities in its environment. Is it interacting with others based on expected expectations? Or, does it exist in a different way? According to this inquiry, they will make a decision on how to treat the entity at stake. This is a necessary step in order for the Ego-X relationship to be generally acknowledged as social relationship. That is, Ego's interpretation of X ("a reliable partner in social interaction") must pass a triadically-structured audit to become a reality beyond the frame of dyadic interaction. If X cannot be identified as such, the interpretation by the first actor (A) turns out to be a misjudgment. His/her interaction with X will be classified as something different than social interaction. In terms of the entity X in Alter-position, she can no longer leave undisturbed her idiosyncratic interpretation on this entity.

(2) Once the problem of another self has been cleared, the Ego-Alter relationship can take a further step. Their interactions enter the stage of stabilization through a triangulation of perspectives. Here again, the third party plays a decisive role. Ego and Alter experience themselves in a You-and-I relationship in front of Tertius, that is, from an objectifying view. Ego experiences itself in a relationship with another self (Alter) from the perspective of Tertius. Ego is aware that it interprets Alter's expectations in the presence of Tertius. Correspondingly, the same is true for Alter. An important consequence of the inclusion of the third actor is that it triangulates the perspectives within the dyadic interaction frame (Lindemann 2005, see also Berger and Luckmann 1991). Absorbing a third person perspective into the process of mutual understanding enables both participants to observe their interaction "from the outside." The closed loop of reciprocal expectations between Ego and Alter is exposed to the eyes of those who co-expect. As a result, the dyadic structure of expectations of expectations is objectified by reference to the assumed co-expectations on the part of Tertius.

Within this triadic structure, the interpretative relationship between Ego and Alter is simultaneously an observed relationship. Since it is an observed relationship, it is possible to distinguish between its current performance and a generalizable pattern that structures the relationship. Binding effects generated through the triangulation of the dyad are not to be attributed to the perspectives of one single party. This is the reason why one can speak of reflexivity in a triadically-structured relationship. The point in question is not a closed interaction between two selves, but a relationship in which those involved can alternately take a position of "I", "you", "she/he", "they", and "we." In this regard, the third actor is a precondition for the formulation of social order.

Each of the three positions is not necessarily a single entity, but can respectively be taken by a collective as well. Another important point is that the physical co-presence of Tertius is, strictly speaking, not always required for the mutual interpretation within a dyadic Ego-Alter interaction (Luhmann 1985). In practice, actor(s) in Tertius-position may be not present on site or even unidentifiable as concrete individuals. In such cases, they exist as those who can potentially be involved in actual communications and, if necessary, be addressed to as such. The relevance of third person perspective becomes apparent when one assumes an external real party who can have influences on the course of a particular dyadic interaction. A teacher (or rather, her perspectives) can affect a dispute between two students even if she is physically not present during the event; this is possible insofar as she is expected to be available at a later time as a mediator who will listen to both sides of the story.

The relational model of above is highly abstract. It is free from anthropological assumptions in that three figures (Ego/Alter/Tertius) are conceptualized merely as structural positions of interaction process. In an actual situation, different types of entities (or groups of entities) can take one position after another. In case of a normative conflict, both Alter and Ego refer to a third actor whose expectations are considered to be critical for the dispute settlement. It is an open question which concrete entities will relate to each other in this way. It is determined from an objectifying view of Tertius who or what has to be recognized as a reliable Alter, i.e., as an authentic counterpart in social interaction. In case the existence of another self becomes questionable, this will be validated by reference to assessment criteria, which are generally accepted within a given society. An actor in Ego-position can justify his interpretation of the entity in question only on the basis of the objectifying view of third actors. This implies that those in Tertius-position will usually be auditing actors who have a considerable say in the decision-making on questionable entities in Alter-position.

Concluding Remarks

From the standpoint of the formal theory discussed above, it remains an open question whether or not autonomous human-like robots will be included into the sphere of legitimate actors of social order. To answer this question it needs to be examined as to whether they will be interpreted and treated in the field as beings that should be understood (or something else). Given the perspective of the modern border regime, it is likely that anthropomorphic robots will be perceived as questionable entities with potential to cause borderline cases. The formal concept of the social outlined above can guide empirical study of such border issues. Sociality is defined in this framework with reference to the specific complexity of interaction among embodied selves, which leads to the formation of a common order. In principle, no concrete entity will be excluded from the outset. The radically deanthropologized criterion allows us to observe how the status of robots is defined in triadically-structured processes of interaction and how the borders of the social are drawn in a real-world context. As seen above, those in Tertius-position will be powerful actors whose perspectives can affect people's decision about how to treat these entities.

Modern functionally-differentiated society can be characterized as a society, where knowledge has become a key resource for economic and social development, and those who are engaged in knowledge-intensive professions have a significant voice in the politically relevant rulemaking procedure (Jasanoff 2006, Stehr and Grundmann 2011). The relevance of expert knowledge is obvious also in terms of the demarcation between persons and nonpersons, which in many present-day societies, corresponds to human/nonhuman distinctions. Those borders are usually drawn on the basis of scientific evidence and rationale. The reference to well-confirmed scientific 'facts' is considered essential to highlight the singularity of human nature in contrast to other beings and thereby to defend the exceptional position of human beings. Moreover, arguments regarding controversial border issues often refer to particular ethico-legal expertise (or religious world views) in order to gain the overall legitimacy—as in the case of the dispute over the status of unborn human fetuses as well as the status of severely brain damaged and comatose individuals. These phenomena suggest that as regards the issue of robot's status too, experts in science, law, and ethics will come into play as third parties, whose insights can exert greater influence on associated decision-making than other participants.

The likelihood of robots as "social *vis-à-vis*" depends on whether their performances in actual everyday interactions will be perceived as indicating their potential for mutual understanding, and whether this interpretation will evolve into an issue on a more general level. While some analyses do not dismiss the idea of attributing full legal personality to robots (e.g. Foerst 2009, Asaro 2012), this possibility is excluded from the in-depth investigation on the ground that "[t]hinking about artificial humans is a premature effort" (euRobotics 2012, p. 63). Nevertheless, there are many robot-related phenomena which can be seen as a sign of this kind of future development. An example would be the perception gap between developers and users (Lindemann and Matsuzaki 2014). Robotics engineers seem quite aware of the fact that their creatures lack some crucial characteristics of what it is that makes a social actor; but this is not true for lay users. An autonomous anthropomorphic robot appears to them often as a magical apparatus or sometimes even as "a post-human entity" (Leroux and Labruto 2012, p. 7). The twofold reality about robots' features and abilities may grow and polarize public opinion on the issues of moral/legal status of robots. Such a conflict can only be resolved in a triadically-structured process of interaction.

References

Adam, A. 2005. "Delegating and Distributing Morality: Can We Inscribe Privacy Protection in a Machine?" *Ethics and Information Technology* 7(4): 233-42. doi: http://dx.doi.org/10.1007/s10676-006-0013-3.

Akrich, M. 1992. "The De-Scription of Technical Objects." In *Shaping Technology/ Building Society. Studies in Sociotechnical Change*, edited by Wiebe E. Bijker and John Law, 205-24. Cambridge/MA: MIT Press.

Alač, M., J. Movellan, and F. Tanaka. 2011. "When a Robot Is Social: Spatial Arrangements and Multimodal Semiotic Engagement in the Practice of Social Robotics." *Social Studies of Science* 41(6): 893-926. doi: http://dx.doi. org/10.1177/0306312711420565.

Allen, C., I. Smit, and W. Wallach. 2005. "Artificial Morality: Top-Down, Bottom-up, and Hybrid Approaches." *Ethics and Information Technology* 7(3): 149-55. doi: http://dx.doi.org/10.1007/s10676-006-0004-4.

Allen, T. and R. Widdison. 1996. "Can Computers Make Contracts?" *Harvard Journal of Law and Technology* 9(1): 25-52.

Andrade, F., P. Novais, J. Machado, and J. Neves. 2007. "Contracting Agents: Legal Personality and Representation." *Artificial Intelligence and Law* 15(4): 357-373. doi: http://dx.doi.org/10.1007/s10506-007-9046-0.

Asaro, P.M. 2012. "A Body to Kick, but Still No Soul to Damn: Legal Perspectives on Robotics." In *Robot Ethics: The Ethical and Social Implications of Robotics*, edited by Patrick Lin, Keith Abney and George Bekey, 169-86. Cambridge, MA: MIT Press.

Beck, S. 2013. "Über Sinn und Unsinn von Statusfragen – Zu Vor- und Nachteilen der Einführung einer elektronischen Person." In *Robotik und Gesetzgebung*, edited by Eric Hilgendorf and Jan-Philipp Günther, 239-60. Baden-Baden: Nomos.

Berger, P.L. and T. Luckmann. 1991. *The Social Construction of Reality: A Treatise in the Sociology of Knowledge*. Harmondsworth: Penguin. Original edition, 1966.

Callon, M. 1986. "The Sociology of an Actor-Network: The Case of the Electric Vehicle." In *Mapping the Dynamics of Science and Technology. Sociology of Science in the Real World*, edited by Michel Callon, John Law and Arie Rip, 19-34. London: Macmillan Press.

Calverley, D.J. 2008. "Imagining a Non-Biological Machine as a Legal Person." *AI & Society* 22(4): 523-37. doi: http://dx.doi.org/10.1007/s00146-007-0092-7.

Cerulo, K.A. 2009. "Nonhumans in Social Interaction." *Annual Review of Sociology* 35: 531-52. doi: http://dx.doi.org/10.1146/annurev-soc-070308-120008.

Coeckelbergh, M. 2010. "Robot Rights? Towards a Social-Relational Justification of Moral Consideration." *Ethics and Information Technology* 12(3): 209-21. doi: http://dx.doi.org/10.1007/s10676-010-9235-5.

Darling, K. 2012. Extending Legal Rights to Social Robots, "We Robot" Conference, Miami, Fl, 21-22 April 2012.

euRobotics. 2012. Suggestion for a Green Paper on Legal Issues in Robotics, Edited by Christophe Leroux and Roberto Labruto. http://www.eu-robotics.net/cms/upload/PDF/euRobotics_Deliverable_D.3.2.1_Annex_ Suggestion_GreenPaper_ELS_IssuesInRobotics.pdf (accessed February 8, 2015): euRobotics.

Evans, E.P. 1987. *The Criminal Prosecution and Capital Punishment of Animals.* London: Faber and Faber. Original edition, 1906.

Feil-Seifer, D., K. Skinner, and M.J. Matarić. 2007. "Benchmarks for Evaluating Socially Assistive Robotics." *Interaction Studies* 8(3): 423-39.

Floridi, L. and J.W. Sanders. 2004. "On the Morality of Artificial Agents." *Minds and Machines* 14(3): 349-79. doi: http://dx.doi.org/10.1023/ b:mind.0000035461.63578.9d.

Foerst, A. 2009. "Robots and Theology." *Erwägen Wissen Ethik* 20(2): 181-93.

Ishiguro, H. and S. Nishio. 2007. "Building Artificial Humans to Understand Humans." *Journal of Artificial Organs* 10(3): 133-42. doi: http://dx.doi. org/10.1007/s10047-007-0381-4.

Jank, M. 2014. *Der Homme Machine Des 21. Jahrhunderts: Von Lebendigen Maschinen Im 18. Jahrhundert Zur Humanoiden Robotik der Gegenwart.* Paderborn: Fink.

Jasanoff, S. 2006. "Ordering Knowledge, Ordering Society." In *States of Knowledge: The Co-Production of Science and the Social Order*, edited by Sheila Jasanoff, 13-45. London & New York: Routledge.

Johnson, D.G. 2006. "Computer Systems: Moral Entities but Not Moral Agents." *Ethics and Information Technology* 8(4): 195-204. doi: http://dx.doi. org/10.1007/s10676-006-9111-5.

Karnow, C.E.A. 1994. "The Encrypted Self: Fleshing out the Rights of Electronic Personalities." *The John Marshall Journal of Computer and Information Law* 13(1): 1-16.

Kelsen, H. 1943. *Society and Nature: A Sociological Inquiry.* Chicago, IL: University of Chicago Press.

Latour, B. 1992. "Where Are the Missing Masses? The Sociology of a Few Mundane Artifacts." In *Shaping Technology/Building Society. Studies in Sociotechnical Change*, edited by Wiebe E. Bijker and John Law, 225-58. Cambridge, MA: MIT Press.

Latour, B. 2005. *Reassembling the Social: An Introduction to Actor-Network-Theory.* Oxford: Oxford University Press.

Leroux, C. and R. Labruto. 2012. "Eurobotics Project – Deliverable D3.2.1 Ethical Legal and Societal Issues in Robotics." Accessed February 7, 2015. http://www. eurobotics-project.eu/cms/upload/PDF/euRobotics_Deliverable_D.3.2.1_ ELS_IssuesInRobotics.pdf.

Lindemann, G. 2005. "The Analysis of the Borders of the Social World: A Challenge for Sociological Theory." *Journal for the Theory of Social Behaviour* 35(1): 69-98. doi: http://dx.doi.org/10.1111/j.0021-8308.2005.00264.x.

Lindemann, G. 2008. "Lebendiger Körper – Technik – Gesellschaft." In *Die Natur der Gesellschaft. Verhandlungen Des 33. Kongresses der Deutschen Gesellschaft Für Soziologie in Kassel 2006*, edited by Karl-Siegbert Rehberg, 689-704. Frankfurt am Main & New York: Camups.

Lindemann, G. 2009a. *Das Soziale Von Seinen Grenzen Her Denken*. Weilerswist: Velbrück Wissenschaft.

Lindemann, G. 2009b. "Gesellschaftliche Grenzregime und soziale Differenzierung." *Zeitschrift für Soziologie* 38(2): 92-110.

Lindemann, G. 2010. "The Lived Human Body from the Perspective of the Shared World (*Mitwelt*)." *Journal of Speculative Philosophy* 24(3): 275-91.

Lindemann, G. 2011. "On Latour's Social Theory and Theory of Society, and His Contribution to Saving the World." *Human Studies* 34(1): 93-110. doi: http:// dx.doi.org/10.1007/s10746-011-9178-9.

Lindemann, G. and H. Matsuzaki. 2014. "Constructing the Robot's Position in Time and Space: The Spatio-Temporal Preconditions of Artificial Social Agenc." *Science, Technology & Innovation Studies* 10(1): 85-106.

Luckmann, T. 1970. "On the Boundaries of the Social World." In *Phenomenology and Social Reality. Essays in Memory of Alfred Schutz*, edited by Maurice Natanson, 73-100. The Hague: Nijhoff.

Luhmann, N. 1965. *Grundrechte Als Institution: Ein Beitrag Zur Politischen Soziologie*. Vol. Bd. 24. Berlin: Duncker & Humblot.

Luhmann, N. 1985. *A Sociological Theory of Law* London: Routledge & Kegan Paul. Original edition, 1972.

Luhmann, N. 1995. *Social Systems*. Stanford, CA: Stanford University Press. Original edition, 1984.

MacDorman, K.F. and H. Ishiguro. 2006. "The Uncanny Advantage of Using Androids in Cognitive and Social Science Research." *Interaction Studies* 7(3): 297-337. doi: http://dx.doi.org/10.1075/is.7.3.03mac.

Matsuzaki, H. 2011. "Die Frage Nach Der „Agency" Von Technik und die Normenvergessenheit der Techniksoziologie." In *Akteur – Individuum – Subjekt: Fragen Zu ‚Personalität' und ‚Sozialität'*, edited by Nico Lüdtke and Hironori Matsuzaki, 301-325. Wiesbaden: VS Verlag für Sozialwissenschaften.

Matthias, A. 2004. "The Responsibility Gap: Ascribing Responsibility for the Actions of Learning Automata." *Ethics and Information Technology* 6(3): 175-83. doi: http://dx.doi.org/10.1007/s10676-004-3422-1.

Matthias, A. 2008. *Automaten als Träger von Rechten. Plädoyer für eine Gesetzesänderung*. Berlin: Logos.

Meister, M. 2014. "When Is a Robot Really Social? An Outline of the Robot Sociologicus." *Science, Technology & Innovation Studies* 10(1): 107-34.

Menciassi, A. and C. Laschi. 2012. "Biorobotics." In *Handbook of Research on Biomedical Engineering Education and Advanced Bioengineering Learning: Interdisciplinary Concepts*, edited by Ziad O. Abu-Faraj, 490-520. Hershey, PA: Medical Information Science Reference.

Nass, C., and Y. Moon. 2000. "Machines and Mindlessness: Social Responses to Computers." *Journal of Social Issues* 56(1): 81-103. doi: http://dx.doi.org/10.1111/0022-4537.00153.

Neumann, A. 2007. Teufelsbund und Teufelspakt (Mittelalter). Lexikon Zur Geschichte Der Hexenverfolgung.

Parsons, T. 1968. "Interaction: I. Social Interaction." In *International Encyclopedia of the Social Sciences*, edited by David L. Sills and Robert K. Merton, 429-41. New York: McGraw-Hill.

Pfadenhauer, M. 2014. "On the Sociality of Social Robots. On the Sociality of Social Robots." *Science, Technology & Innovation Studies* 10(1): 135-53.

Plessner, H. 1975. *Die Stufen des Organischen und der Mensch. Einleitung in die philosophische Anthropologie*. Berlin: de Gruyter. Original edition, 1928.

Powers, T.M. 2011. "Incremental Machine Ethics: Adaptation of Programmed Constraints." *IEEE Robotics & Automation Magazine* 18(1): 51-8. doi: http://dx.doi.org/10.1109/MRA.2010.940152.

Rammert, W. 2012. "Distributed Agency and Advanced Technology, Or: How to Analyze Constellations of Collective Inter-Agency." In *Agency without Actors? New Approaches to Collective Action*, edited by Jan-Hendrik Passoth, Birgit Peuker and Michael Schillmeier, 89-112. London: Routledge.

Rammert, W. and I. Schulz-Schaeffer. 2002. Technik und Handeln: Wenn soziales Handeln sich auf menschliches Verhalten und technische Artefakte verteilt. In *TUTS Working Paper*. Berlin: Technische Universität Berlin - Institut für Soziologie.

Schütz, A. 1974. *Der sinnhafte Aufbau der sozialen Welt: Eine Einführung in die verstehende Soziologie*. Frankfurt am Main: Suhrkamp. Original edition, 1932.

Simmel, G. 1911. "How Is Society Possible?" *American Journal of Sociology* 16(3): 372-91. Original edition, 1910.

Simmel, G. 1992. *Soziologie: Untersuchungen über die Formen der Vergesellschaftung, Gesamtausgabe Bd. 11*. Frankfurt am Main: Suhrkamp. Original edition, 1908.

Solum, L.B. 1992. "Legal Personhood for Artificial Intelligences." *North Carolina Law Review* 70(4): 1231-87.

Sparrow, R. 2007. "Killer Robots." *Journal of Applied Philosophy* 24(1): 62-77. doi: http://dx.doi.org/10.1111/j.1468-5930.2007.00346.x.

Stahl, B.C. 2006. "Responsible Computers? A Case for Ascribing Quasi-Responsibility to Computers Independent of Personhood or Agency." *Ethics and Information Technology* 8(4): 205-13. doi: http://dx.doi.org/10.1007/s10676-006-9112-4.

Stehr, N. and R. Grundmann. 2011. *Experts: The Knowledge and Power of Expertise*. London: Routledge.

Teubner, G. 2006. "Rights of Non-Humans? Electronic Agents and Animals as New Actors in Politics and Law." *Journal of Law and Society* 33(4): 497-521. doi: http://dx.doi.org/10.1111/j.1467-6478.2006.00368.x.

Tonkens, R. 2009. "A Challenge for Machine Ethics." *Minds and Machines* 19(3): 421-38. doi: http://dx.doi.org/10.1007/s11023-009-9159-1.

Torrance, S. 2008. "Ethics and Consciousness in Artificial Agents." *AI & Society* 22(4): 495-521. doi: http://dx.doi.org/10.1007/s00146-007-0091-8.

United Nations. 1948. "The Universal Declaration of Human Rights." United Nations Accessed March 10, 2015. http://www.un.org/en/documents/udhr.

Verbeek, P-P. 2006. "Materializing Morality: Design Ethics and Technological Mediation." *Science, Technology & Human Values* 31(3): 361-380. doi: http://dx.doi.org/10.1177/0162243905285847.

Verbeek, P-P. 2009. "Ambient Intelligence and Persuasive Technology: The Blurring Boundaries between Human and Technology." *NanoEthics* 3(3): 231-42. doi: http://dx.doi.org/10.1007/s11569-009-0077-8.

Verschraegen, G. 2002. "Human Rights and Modern Society: A Sociological Analysis from the Perspective of Systems Theory." *Journal of Law and Society* 29(2): 258-81. doi: http://dx.doi.org/10.1111/1467-6478.00218.

Wallach, W. and C. Allen. 2009. *Moral Machines: Teaching Robots Right from Wrong*. Oxford: Oxford University Press.

Weber, M. 1946. "Science as a Vocation." In *From Max Weber: Essays in Sociology*, edited by Hans Heinrich Gerth and Charles Wright Mills, 129-56. New York: Oxford University Press. Original edition, 1922.

Weber, M. 1980. *Wirtschaft und Gesellschaft. Grundriß der verstehenden Soziologie*. Tübingen: Mohr. Original edition, 1921-22.

Wettig, S. and E. Zehendner. 2004. "A Legal Analysis of Human and Electronic Agents." *Artificial Intelligence and Law* 12(1-2): 111-35. doi: http://dx.doi.org/10.1007/s10506-004-0815-8.

Yamazaki, R., S. Nishio, H. Ishiguro, M. Nørskov, N. Ishiguro, and G. Balistreri. 2012. "Social Acceptance of a Teleoperated Android: Field Study on Elderly's Engagement with an Embodied Communication Medium in Denmark." *Lecture Notes in Computer Science* 7621: 428-37. doi: http://dx.doi.org/10.1007/978-3-642-34103-8_43.

Zhao, S. 2006. "Humanoid Social Robots as a Medium of Communication." *New Media & Society* 8(3): 401-19. doi: http://dx.doi.org/10.1177/1461444806061951.

Chapter 10

The Diffuse Intelligent Other: An Ontology of Nonlocalizable Robots as Moral and Legal Actors

Matthew E. Gladden

Much thought has been given to the question of who bears moral and legal responsibility for actions performed by robots. Some argue that responsibility could be attributed to a robot if it possessed human-like autonomy and metavolitionality, and that while such capacities can potentially be possessed by a robot with a single spatially compact body, they cannot be possessed by a spatially disjunct, decentralized collective such as a robotic swarm or network. However, advances in ubiquitous robotics and distributed computing open the door to a new form of robotic entity that possesses a unitary intelligence, despite the fact that its cognitive processes are not confined within a single spatially compact, persistent, identifiable body. Such a "nonlocalizable" robot may possess a body whose myriad components interact with one another at a distance and which is continuously transforming as components join and leave the body. Here we develop an ontology for classifying such robots on the basis of their autonomy, volitionality, and localizability. Using this ontology, we explore the extent to which nonlocalizable robots—including those possessing cognitive abilities that match or exceed those of human beings—can be considered moral and legal actors that are responsible for their own actions.

Introduction

Philosophers, roboticists, and legal scholars have given much thought to the challenges that arise when attempting to assign moral and legal responsibility for actions performed by robots. One difficulty results from the fact that the word "robot" does not describe a single species or genus of related beings, but rather a vast and bewildering universe of entities possessing widely different morphologies and manners of functioning.

To date, scholars exploring questions of moral and legal responsibility have largely focused on two types of robots. One kind comprises telepresence robots that are remotely operated by a human being to carry out tasks such as performing surgery, giving a lecture, or firing a missile from an aerial vehicle

(Datteri 2013, Hellström 2012, Coeckelbergh 2011). Such robots are not morally or legally responsible for their own actions; instead, responsibility for their actions is attributed to their operators, designers, or manufacturers according to well-established legal and moral frameworks relating to human beings' use of tools and technology. It is possible for such a robot to be designed and manufactured in one country, relocated to (and acting in) a second country, and remotely controlled by a human operator based in a third country, thus raising questions about which nation's laws should be used to assign responsibility for the robot's actions. However, in principle it is relatively easy to trace the chain of causality and identify the locations in which each stage of a robot's action occurred.

The other main type of robot for whose actions scholars have sought to attribute moral and legal responsibility comprises autonomous devices such as self-driving cars (Kirkpatrick 2013) and autonomous battlefield robots (Sparrow 2007). Here questions of legal and moral responsibility are complicated by the fact that it is not immediately obvious who—if anyone—has "made the decision" for the robot to act in a particular way. Depending on the circumstances, arguments can be made for attributing responsibility for a robot's action to the robot's programmer, manufacturer, owner, or—if the robot possesses certain kinds of cognitive properties—even to the robot itself (Sparrow 2007, Dreier and Spiecker genannt Döhmann 2012, p. 211). At the same time, the attribution of legal responsibility for the robot's actions is simplified by the fact that the computational processes guiding an autonomous robot's behavior typically take place within the robot's own spatially compact body, reducing the likelihood that the robot's process of acting could cross national borders.

While efforts to account for the actions of some kinds of robots are thus already well-advanced, at the frontiers of ubiquitous robotics, nanorobotics, distributed computing, and artificial intelligence, experts are pursuing the development of a new form of robotic entity whose manner of being, deciding, and acting is so radically different from those of earlier robots that it will be difficult or impossible to apply our traditional conceptual frameworks when analyzing whether such a robot is morally and legally responsible for its actions. This new kind of being is the *nonlocalizable* robot, one whose cognitive processes do not subsist within a single identifiable body that is confined to a particular set of locations and that endures across time (Yampolskiy and Fox 2012, pp. 129-38, Gladden 2014b, p. 338). Such a robot might take the form, for example, of a vast digital-physical ecosystem in which millions of interconnected devices participate in a shared cognitive process of reaching decisions and acting, even as devices continually join and leave the network; or it could exist as a loosely-coupled, free-floating oceanic cloud of nanorobotic components that communicate with one another via electromagnetic signals while drifting among the world's seas; or it could take the form of an evolvable computer virus whose cognitive processes are hidden within an ever-shifting network computers around the world—ones found not only in homes and business but also in airplanes, ships, and orbiting satellites.

Attempts to address questions of responsibility for actions performed by nonlocalizable robots raise a multitude of complex philosophical and legal questions that have not yet been carefully explored. For example, in an environment containing such robots it might be apparent that "some" robotic entity has just acted, but it may be impossible to correlate that action with just a single robot, as a networked device can be part of the bodies of many different nonlocalizable robots simultaneously, and over the span of a few milliseconds or minutes a networked device might undergo the processes of becoming integrated into a nonlocalizable robot's body, acting as a part of its body, and then becoming dissociated from its body. More fundamentally, it may even be impossible to determine with clarity that a particular robot *exists*—to identify it and distinguish it from other entities and delineate the boundaries of its physical and cognitive existence. The fact that such a spatially diffuse (and potentially even globally extensive) robotic body might someday be occupied by an artificial intelligence possessing human-like cognitive capacities (Yampolskiy and Fox 2012, pp. 133-8) adds another dimension that one must account for when attempting to develop a framework for determining the moral and legal responsibility that nonlocalizable robots might bear for their actions.

In this text we suggest the outlines for such a framework. We begin in the following section by proposing an ontology of autonomy, volitionality, and localizability that will allow us to describe essential physical and cognitive characteristics of nonlocalizable robots and analyze the extent to which they are morally and legally responsible actors.[1]

Developing an Ontological Framework

Purpose of the Ontology

There have been efforts by computer scientists to develop a universal "ontology" for robotics that defines its current terminology and engineering principles in a way that is "formally specified in a machine-readable language, such as first-order logic" (Prestes et al. 2013, p. 1194). Such specialized technical schemas facilitate the standardization of robotics engineering and interoperability of different robotic systems, however they do not attempt to delineate the full universe of ways of being and acting that are available for robots and which philosophers can use to

---

1 The word "actors" is used here to mean "entities who are morally or legally responsible for their own actions." Use of the word "agent" has been avoided, due to the fact that it possesses different (and in many ways incompatible) meanings when used in the contexts of moral philosophy, law, and computer science. Similarly, the posing of the question of whether a nonlocalizable robot can be considered a moral or legal "person" has generally been avoided, since—depending on the context—personhood does not necessarily imply that an entity is morally and legally responsible for its actions.

explore the social, ethical, and legal questions that are provoked by the existence of ever more sophisticated robotic morphologies.

A more robust ontology should be capable of describing a robotic entity both at its most fundamental level of physical reality as well as at the level of emergent phenomena that are attributed to the entity as a result of higher-level interactions with its environment (Gladden 2014b, p. 338). Such an ontology would allow us to analyze a robot from the perspective of its nature as an autonomous viable system that organizes matter, energy, and information (Gladden 2014a, p. 417), to its form and functioning as a concrete physical device that incorporates sensors, actuators, and computational processors, to its role within human (or artificial) societies as a social, political, legal, and cultural object or subject.

Here we focus on one element of such an ontology that poses a significant—and thus far largely unexplored—challenge to the current debate on whether robots can bear moral and legal responsibility for their actions. Namely, we introduce the concept of a robot's *localizability*, and in particular we will consider the question of moral and legal responsibility for the actions of a robot that is a nonlocalizable being—i.e., one that has a physical body, but whose body does not exist in any one particular place across time. Our ontology also encompasses two other relevant elements: those of a robot's levels of *autonomy* and *volitionality*. A brief discussion of these three concepts follows.

Autonomy

Autonomy relates to an entity's ability to act without being controlled. For robots, possessing autonomy means being "capable of operating in the real-world environment without any form of external control for extended periods of time" (Bekey 2005, p. 1). In its fullest form, autonomy involves not only performing cognitive tasks such as setting goals and making decisions but also performing physical activities such as securing energy sources and carrying out self-repair without human intervention. Building on conventional classifications of robotic autonomy (Murphy 2000, pp. 31-4), we can say that currently existing robots are either *nonautonomous* (e.g., telepresence robots that are fully controlled by their human operators when fulfilling their intended purpose, or robots which do not act to fulfill any purpose), *semiautonomous* (e.g., robots that require "continuous assistance" or "shared control" in order to fulfill their intended purpose), or *autonomous* (e.g., robots that require no human guidance or intervention in fulfilling their intended purpose). We can use the term *superautonomous* to describe future robots whose degree of autonomy may significantly exceed that displayed by human beings—e.g., because the robots' ability to independently acquire new knowledge frees them from any need to seek guidance from with human subject-matter experts or because their bodies contain an energy source that can power them throughout their anticipated lifespan.

Volitionality

Volitionality relates to an entity's ability to self-reflexively shape the intentions that guide its actions. A robot is *nonvolitional* when it possesses no internal goals or "desires" for achieving particular outcomes nor any expectations or "beliefs" about how performing certain actions would lead to particular outcomes. Some telepresence robots are nonvolitional, insofar as a human operator supplies all of the desires and expectations for their actions; the robot is simply a pliable, transparent tool. A robot is *semivolitional* if it possesses *either* a goal of achieving some outcome *or* an expectation that particular outcomes will result if the robot acts in a certain way, but the robot does not link goals with expectations. For example, such a robot might have been programmed with the goal of "moving to the other side of the room," but it has no means of interpreting the sensory data that it is receiving from its environment to know where it is, nor does it have an understanding of the fact that activating its actuators would cause it to "move." A robot is *volitional* if it combines goals with expectations; in other words, it can possess an intention,[2] which is a mental state that comprises both a desire and a belief about how some act that the robot is about to perform can contribute to fulfilling that desire (Calverley 2008, p. 529). For example, a therapeutic social robot might have a goal of evoking a positive emotional response in its human user, and its programming tells it that by following particular strategies for social interaction it is likely to evoke such a response.

We can describe as *metavolitional* a robot that possesses what scholars have elsewhere referred to as a "second-order volition," or an intention *about* an intention (Calverley 2008, pp. 533-5). For example, imagine a more sophisticated therapeutic social robot that comes to realize that it is "manipulating" its human users by employing subtle psychological techniques to coax them into displaying positive emotional responses. Such a robot would display metavolitionality if it grew weary of what it now saw as its "dishonest" and coercive behavior, and it wished that it did not feel compelled to constantly manipulate human beings' psyches. Metavolitionality is a form of volitionality typically demonstrated by adult human beings. We can use the term *supervolitional* to describe a possible future robot that regularly forms intentions that are higher than second-order. For example, such a robot might possess a mind that is capable of simultaneously experiencing thousands of different second-order intentions and using a third-order intention to guide and transform those second-order intentions in a concerted way that reshapes the robot's character.

---

2   When used in this text, "intentionality" is employed in the usual philosophical sense to describe an entity's ability to possess mental states that are directed toward (or "about") some object; that is a broader phenomenon than the possession of a particular "intention" as defined here.

Localizability

Localizability relates to the extent to which an entity possesses a stable, identifiable physical body that is confined to one or more concrete locations. A robot is *local* when its "sensing body" of environmental sensors, its "acting body" of actuators and manipulable physical components, and the "brain" in which its cognitive processes are executed are found together in a single location and possess physical forms that are discrete and easily identifiable and endure over time. Such robots might include a robotic vacuum cleaner, an articulated industrial robot controlled by a teach pendant that is connected to the robotic arm by a physical cable, or an experimental wheeled robot whose body moves through a maze while the its data analysis and decision-making are being carried out in a dedicated desktop computer that is located in the same room as the robot's sensing and acting body and is wirelessly linked to it.

A robot is *multilocal* when it comprises two or more stable and clearly identifiable components that together form the robotic unit and which are not in physical proximity to one another. These components can potentially be located in different parts of the world: for example, a telesurgery robot might include a surgical device located in a hospital room and a controller unit that is manipulated by a human surgeon in a different country and is linked to the surgical device through the Internet. Similarly, an autonomous battlefield robot system could include a central computer located in a military base that is simultaneously controlling the actions of several robotic bodies participating in a military operation in a conflict zone in another country. A robotic entity can be "intensely multilocal" if it possesses a very large number of components that are located at great distances from one another, however this is a quantitative rather than qualitative distinction: as long as a robotic entity's physical components are identifiable and endure over time, the robot is still multilocal, regardless of how numerous or widely dispatched the components might be. Many current and anticipated systems for ambient intelligence and ubiquitous robotics are multilocal (Défago 2001, Pagallo 2013, Weber and Weber 2010).

A robotic entity is truly *nonlocalizable* when it is impossible to clearly specify in any given moment the exact location or extent of the robot's body or to specify exactly "where" key activities of the robot's cognitive processes are taking place. It may be known that the robot exists, and from the nature of the robot's actions in the world it is known that the robot must possess both physical sensors and actuators and some substrate upon which its cognitive processes are being executed, however it is not possible to specify exactly where or in what form all of those components exist.

The Sources of Nonlocalizability

Nonlocalizability can result from a number of conditions. For example, at any given moment, a robot's physical extension may include components that are a part of its body, components that are in the process of being removed from its body, and components that are in the process of becoming part of its body. If the robot is able to add new components to its body and remove old ones from its body at a sufficient rate, it is possible that the robot's body at one moment in time might not share any physical components in common with the robot's body as it existed just a few moments earlier.[3] Such a robot is continuously in the act of moving its cognitive processes into a new body. In this case it is not the physical components of its body that give the robot its identity, insofar as those components are in the perpetual process of being discarded and replaced with new ones; rather it is the relationship between its components—its organizing principle that arranges matter, energy, and information into an enduring viable system (Gladden 2014a, p. 417)—that gives a robot its identity as a potential moral and legal actor.[4] Another source of nonlocalizability occurs if a robot's primary cognitive process consists of a unitary neural computing process that can be distributed across a cluster or grid of processors rather than a program that is executed on a single serial processor (Gladden 2014b, p. 338). Finally, nonlocalizability can also result from the fact that a single networked device such as a sensor or actuator can potentially be part of the "bodies" of several different robots simultaneously, producing a situation in which robots' bodies may partly overlap with one another, complicating the question of specifying "to which robot" a particular component or action belongs.

Note that as they currently exist, conventional robotic swarms and networks may be intensely multilocal, but they are not nonlocalizable. While the miniaturization of parts and multiplication of parts constituting a robot's body might contribute to nonlocalizability, they are insufficient to create it, insofar as a robot can have a body consisting of many small parts that is still identifiable and stable over time. Moreover, if the components of a swarm lack a centralized cognitive process that controls their actions, then they are better understood as a collection of independent robots than as a single robot whose body comprises numerous parts. A nonlocalizable robot is not simply a multi-agent system or collection of autonomous robots, but a single entity possessing a unitary identity that can be at least theoretically capable of possessing autonomy and metavolitionality.

---

3 Of course, human bodies also undergo a process of gradually replacing their components; however in the case of a nonlocalizable robot, this becomes significant insofar as the transformation of the robot's body is taking place on a time-scale relevant to the time-frame (Murphy 2000, pp. 31-4) in which decisions and actions are made that are the subject of discussions about potential moral and legal responsibility.

4 The debate over whether a robot's identity derives from its physical components or some emergent process at work within them parallels, in some ways, discussions and concepts as ancient as Aristotle's notion of the soul as the form that animates a living substance, the Neoplatonic concept of ψυχή, and the understanding of the λόγος possessed by individual human beings that was developed by the Greek Fathers of the Catholic Church.

The Current Debate over Moral and Legal Responsibility for Robotic Actors

Combining our ontology's four possible values for a robot's level of autonomy, five possibilities for volitionality, and three possibilities for localizability yields a total of 60 prototypical kinds of robotic entities described by the ontology. For some of these robots, the question of moral and legal responsibility for their actions can be analyzed using traditional frameworks that are regularly applied to the actions of human beings. Below we consider several such cases.

The Robot as a Tool for Human Use

Consider once again a multilocal telesurgery robot that consists of a mechanized surgical instrument that operates on a patient and is guided by a human surgeon—who may be located in a different country—manipulating a controller connected to the instrument through the Internet. If the surgeon must directly and specifically request each movement of the surgical instrument, the device would likely be nonautonomous and nonvolitional. If the human surgeon can give the instrument more generalized instructions that it then interprets and executes (such as "Withdraw the laparoscope"), then the robot is operating under a form of "continuous assistance" or "shared control" (Murphy 2000, pp. 31-4), and it may qualify as semiautonomous and semivolitional or even volitional.

As a tool for use by human beings, questions of legal responsibility for any harmful actions performed by such a robot revolve around well-established questions of product liability for design defects (Calverley 2008, p. 533, Datteri 2013) on the part of its producer, professional malpractice on the part of its human operator, and, at a more generalized level, political responsibility for those legislative and licensing bodies that allowed such devices to be created and used. The international dimension of having a human operator who causes the robot to act in a different country raises questions about legal jurisdiction, conflicts of national law, and extraterritoriality (Doarn and Moses 2011), but those issues can be addressed using existing legal mechanisms.

Questions of the moral responsibility for the robot's actions can be similarly resolved on the basis of its functioning as a passive instrument produced and used by human beings for their own ends. Such a robot does not possess responsibility of its own for its actions (Hellström 2012, p. 104), but using Stahl's (2006, p. 210) formulation could be seen as possessing a sort of "quasi-responsibility" that serves as a placeholder to point back to the robot's human operators and producers, who are ultimately responsible for its actions.

The Robot as Ersatz Human Being

Imagine a humanoid social robot that has been designed in such a way that it possesses not only a human-like local body, but also a human-like form of autonomy and metavolitionality. While such a robot might appear—by today's standards—to represent a quite futuristic technological breakthrough, the question of moral and legal responsibility for the robot's actions could largely be addressed using traditional conceptual frameworks.

In the eyes of the law, a robot that is both autonomous and metavolitional can be considered a legal person that is responsible for its actions (Calverley 2008, pp. 534-5). From the legal perspective, "autonomy" means that there is not some human being (or other robot) to whom a robot defers and who controls the robot's decision-making (ibid., p. 532). From the legal perspective, this does not require the robot to possess "free will"; the robot's behavior can indeed be determined by certain preferences or desires, as long as those preferences have somehow been generated from within by the robot itself (ibid., p. 531). A robot whose actions are being directed via remote control—or which has been programmed in advance to act in a specific way in response to a specific situation—would not be acting autonomously. However, autonomy is not, in itself, sufficient to generate legal responsibility; an entity must also act with metavolitionality.

The question of whether such a robot is morally (and not just legally) responsible for its actions is more complex, although it also builds on traditional understandings of autonomy and metavolitionality. While incompatibilists argue that an entity must have the freedom to choose between multiple alternatives in order to be held morally responsible, others like Frankfurt claim that such freedom is unnecessary and that an entity can be morally responsible if it is capable of experiencing not only desires but also desires *about* its desires, as is seen in our human capacity for "changing desires through the sheer force of mental effort applied in a self-reflexive way" (ibid., pp. 531-2). Drawing on Aristotle, Hellström notes that one way of analyzing whether an entity bears moral responsibility for some deed is to ask whether it is "worthy of praise or blame for having performed the action (Hellström 2012, p. 102); if we cannot imagine ourselves expressing praise or blame for a robot's action, then it is likely not the sort of actor to which moral responsibility can be attributed. This approach can be understood as a sort of intuitive "shortcut" for considering whether an entity is metavolitional. Kuflik (1999, p. 174) gets at the idea of metavolitionality similarly, arguing that it only makes sense to assign computer-based devices like robots responsibility for their actions if we are able to ask them:

> ... to provide good reason for their comportment, to assess the force of reasons they had not previously considered, to be willing in some cases to acknowledge the insufficiency of their own reasons and the greater force of reasons not previously considered, to explain mitigating factors and ask for forgiveness, and—failing a show either of good reason or good excuse—to apologize and look for ways of making amends. (ibid., p. 174)

It should be noted, though, that "assignment of responsibility not necessarily is a zero-sum game" (Hellström 2012, p. 104); the fact that a robot bears moral or legal responsibility for its actions does not imply that a different sort of moral and legal responsibility cannot also be attributed to those who brought the robot into existence, educated it in a particular way, or caused it to be placed in the situation in which it was acting.

Swarms and Networks as Localizable Individuals or
Nonlocalizable Communities

Advances in the creation of distributed robotic systems such as networks and swarms of robots (and particularly, swarms of miniaturized robots) have become a topic much studied by philosophers and legal scholars. The rise of such sophisticated robotic systems would seem to point toward the eventual development of robots that are not only autonomous and metavolitional but also truly nonlocalizable; however, to date the development of robotic swarms and networks has not yet spurred significant consideration of potentially autonomous metavolitional nonlocalizable robots as moral or legal actors. Instead, the focus of the debate has moved in other directions.

Some scholars have considered the responsibility of robots that "are connected to a networked repository on the internet that allows such machines to share the information required for object recognition, navigation and task completion in the real world" (Pagallo 2013, p. 501). However, if such a centralized, cloud-based repository is *controlling* the networked robots' actions, then the existence of a stable, identifiable "brain" for the system means that it is at most intensely multilocal; it is not nonlocalizable. On the other hand, if the centralized repository is *not* controlling the individual robots' actions, then the each robot could potentially possess autonomy and metavolitionality, but the system as a whole does not. In that case, the network could be seen as a community of many different robots, a sort of a multi-agent system (Murphy 2000, pp. 293-314). The "membership" of the community might indeed change from moment to moment, as robots join and leave the network. However, the community is not an entity with a single shared cognitive process and metavolition. It more closely resembles a "legal person" such as nation or corporation that can bear legal responsibility for its actions but cannot possess the metavolitionality needed for moral responsibility: a corporation can be said to feel pride or regret only in a metaphorical sense, as the corporation's decisions and actions are made not by some autonomous corporate "mind" but by the minds of its human constituents, often working together (Stahl 2006, p. 210). Building on Ricoeur's notion of "quasi-agency" to refer to the action of nations, Stahl (ibid.) attributes "quasi-responsibility" to a robotic entity that appears to possess moral responsibility but actually does not because it lacks autonomy or metavolitionality; quasi-responsibility serves only as a placeholder that points toward the entities (such as a network's autonomous components) that actually bear responsibility for the system's apparently collective action. In none of these analyses do scholars assert that a robotic entity such as a continuously evolving network could be nonlocalizable while at the same time possessing autonomy and metavolitionality.

Coeckelbergh (2011, pp. 274-5) goes further: rather than arguing that it is impossible for a robotic swarm or network to be both nonlocalizable, autonomous, and metavolitional, he argues that it is not even possible for a "robotic" swarm or network to *exist*. Every robot is inextricably situated within a social and technological ecosystem that includes relations with human beings and other devices, and these relations shape the robot's actions and being. What appears, at first glance, to be a purely robotic network or swarm "can hardly be called a 'robotic' swarm given the involvement of various kinds of systems and humans" that in some sense participate in its decision-making and acting (ibid., p. 274). Thus we see that the debate about robots' possible moral and legal responsibility for their own actions has not yet directly addressed the case of robots that are both autonomous and metavolitional while existing in a physical form that is nonlocalizable. In the following sections we attempt to suggest what such an analysis might look like.

The Future of Moral and Legal Responsibility for Nonlocalizable Robots

Advances Contributing to the Development of Nonlocalizable Robots

Developments in fields such as ubiquitous computing, cooperative nanorobotics, and artificial life are laying the groundwork for the existence of robotic entities whose bodies are capable of shifting from one physical substrate to another, occupying parts of the bodies of other robotic entities, and possessing a spatial extension that comprises many disjunct elements and is not identifiable to human observers—in other words, robotic entities that are truly nonlocalizable. For example, as a "next logical step" that builds on principles of mobile and ubiquitous computing, Défago (2001, pp. 50-53) develops a model of "cooperative robotics" in which teams of nanorobots can coordinate their activities using *ad hoc* wireless networks or other forms of remote physical interaction. We can also anticipate the development of artificial life-forms resembling sophisticated computer viruses which, in effect, are continually exchanging their old bodies for new ones as they move through networked ecosystems to occupy an ever-shifting array of devices in search of "resources for their own survival, growth, and autonomously chosen pursuits" (Gladden 2014a, p. 418). Given the vast spectrum of possibilities unlocked by ubiquitous computing and nanotechnology, it is possible that without even realizing it, human beings could someday find ourselves living inside the "bodies" of nonlocalizable robots that surround us—or their bodies could exist inside of us.

Such nonlocalizable robots might demonstrate a range of possible cognitive capacities similar to those available for localizable robots. Insofar as the architecture utilized by the human brain likely represents only one of many possible substrates for a sapient mind, it seems possible that a single mind that is embodied across multiple, changing, spatially disjunct components could potentially possess levels of autonomy and volitionality at least as great as those of a human being (Yampolskiy and Fox 2012, pp. 133-8). While certain challenges relating to time, space, and computer processing speeds (Gunther 2005, pp. 6-7) arise when attempting to develop artificial general intelligences whose primary cognitive process occurs across a network of spatially disjunct artificial neurons, such spatial dispersion is not likely to prove an insurmountable obstacle to the creation of artificial general intelligence (Loosemore and Goertzel 2012, pp. 93-5). Already work is underway on designing artificial networks (ANNs) whose neurons communicate with one another wirelessly, allowing for the creation of "cyber-physical systems" (CSPs) whose highly flexible and easily expandable system include a "large number of embedded devices (such as sensors, actuators, and controllers) distributed over a vast geographical area" (Ren and Xu 2014, p. 2). Such wireless neural networks could potentially demonstrate capacities impossible for the human brain, insofar as a neuron in such a network is no longer limited to interacting only with neurons that are physically adjacent in three-dimensional space but can potentially connect with other (more distant) neurons to create networks possessing different topologies and dimensionality.

Utilizing our ontological framework, we can now consider the moral and legal responsibility that would be borne for their actions by several types of nonlocalizable robots, beginning with those that demonstrate the lowest levels of autonomy and volitionality.

The Robot as Ambient Magic

Consider a robot whose primary cognitive process can be spread across a vast number of disparate networked computerized devices (Gladden 2014b, p. 338) and which over time shifts elements of that process out of its current devices and into new ones. Such a robot's "body" would consist of all those devices in which its cognitive processes were being executed (or were at least stored) at a given moment in time. By copying parts of its cognitive processes into adjacent networked devices and deleting them from some of those where they had been residing, the robot could in effect "move" in an amoeba-like manner, floating not in the air nor through the ocean's waters but through the global ecosystem of networked devices. It could be massively embodied but in such a way that no one element of its "body" was essential.

If such a nonlocalizable robot were nonautonomous and nonvolitional, it could in some respects resemble a natural phenomenon such as a flickering flame or flowing stream: when engaged physically or digitally by actors such as human beings, it might react in potentially interesting and dramatic ways that display discernible patterns—but not volition. Such a robot might be seen by human beings as a "force of nature" or perhaps even a latent magical energy, a sort of *qi* or mana that is immanent in the environment and can be manipulated by human adepts to produce particular effects (Clarke 1973, p. 36). Such a robot would possess neither legal nor moral responsibility for its actions, but could potentially possess a quasi-responsibility (Stahl 2006, p. 210) that points toward its creators, assuming that it had been purposefully created and had not evolved naturally "in the wild" through the interaction and reproduction of other robotic entities.

The Robot as Diffuse Animal Other

If a robot of the sort just described were able to experience desires and expectations and to purposefully shift its cognitive processing into certain kinds of networked computerized hosts that it found attractive (e.g., those with greater processing power or particular kinds of sensors and actuators) and out of those that were less attractive, it would more closely resemble an animal than an impersonal force. It would no longer move through the Internet of Things randomly—or as steered by external forces—but would proactively explore its environment in a search for those informational, value-storing, agent, or material resources (Gladden 2014a, p. 417) that contribute to the satisfying of its drives for self-preservation and reproduction (Omohundro 2012).

Such a robot could in many ways be considered a sort of digital-physical "animal" with a capacity for bearing moral and legal responsibility similar to those of natural organic animals. (And indeed, in some cases the robot itself might be an organic being (Pearce 2012).) While such robots would not be moral actors responsible for their own actions, in contrast to the "robot as ambient magic" they could potentially be the sort of moral patients to whom moral consideration is due, if—like many kinds of animals—they are able to experience physical or psychological suffering (Gruen 2014).

The Robot as Diffuse Human-like Other

If we now imagine the robot described above to possess human-like cognitive capacities for reasoning, emotion, sociality, and intentionality, it becomes autonomous and metavolitional. With regard to its possession of moral and legal responsibility, it is in some ways similar to the "robot as ersatz human being" mentioned above, however its nonlocalizability means that it lacks some traits that are found in human beings and possesses other characteristics that are impossible for them.

By using its dispersed, networked body to manipulate an environment, such a robot might appear to materialize, ghost-like, from out of the digital-physical ecosystem to speak with, look at, and touch a human being, and then vanish—and in the next moment it could be having a similar social interaction with another human being in another part of the globe. Because of the limitations of our organic bodies, a human being can only be physically located in one place at a time, although we can use technologies like the telephone or videoconference or virtual reality to interact in a mediated fashion with environments in other countries.

However, a nonlocalizable robot could truly be "in" many different locations at once; there is no principle of primacy that allows us to say that one venue of interaction is home to the robot's "real" body and that it is simply projecting itself virtually into the other locations through a "mediating technology"—because the only such technology is the robot's body itself. The fact that the robot is located "here" at a given moment in time does not mean that it is not also somewhere else.

Difficulties arise in attempting to analyze such a robot's possibility for bearing responsibility for its actions. In principle, it might seem capable of bearing human-like responsibility. However, our human ability to bear moral responsibility depends on our metavolitional ability to have *expectations:* an intention requires both a desire and a belief about the outcomes that will result from an action. A being whose body is so radically different from our own may not experience the same sorts of beliefs or expectations about how its environment will be affected by its actions—since the very concepts of "environment" and "action" could have vastly different meanings for such an entity. Similarly, the question of legal responsibility is clouded by the need to determine to which nation's laws the robot is subject, when its body is potentially scattered across continents and seas and satellites in orbit. Identifying "where" in the robot's body a particular decision was made may be as difficult as pinpointing which neurons in a human brain generated a particular action (Bishop 2012, p. 97). Moreover, it is unclear whether the application of traditional legal "rewards" and "punishments" would have meaning for such a nonlocalizable entity.

The Robot as Diffuse Alien Intelligence

Finally we can consider a nonlocalizable robot whose cognitive processes demonstrate superautonomy and supervolitionality. In principle, one might be inclined to impute legal and moral responsibility to such a robot because it possesses levels of autonomy and volitionality that are at least as great as those of human beings. But that overlooks the fact that the robot's experience of its own superautonomy and supervolitionality likely has little in common with our human experiences of our vastly less sophisticated autonomy and metavolitionality. Yampolskiy and Fox (2012, p. 129) argue that "humanity occupies only a tiny portion of the design space of possible minds" and that "the mental architectures and goals of future superintelligences need not have most of the properties of human minds." We might be able to discern from observing such robots' interactions with one another that their actions follow some highly complex and beautiful (or perhaps terrifying) patterns, but it may be impossible for us to determine what portion of those patterns results from a sort of universal natural law that is reflected in the robots' moral sentiments, how much of it is due to legal frameworks that the robots have constructed among themselves (Michaud 2007, p. 243), how much of it is due to their cultural traditions, and how much of it results simply from engineering requirements, mathematical or logical principles, or physical laws. We might never come to understand the internal processes that shape such robots' thoughts, decisions, and actions, because they lack the ability or desire to explain them to us using language or concepts that we can fathom (Abrams 2004). Any attempt by the robots to explain their moral, social, and legal frameworks might require them to employ metaphorical imagery and "overload" our senses in a way that bears more resemblance to the phenomenon of divine revelation than to any processes of logic and reasoning available to the human mind (Gladden 2014b, p. 337).

Serious problems ensue from attempting to apply human notions of "law" to such alien entities. The law "sets parameters, which, as society has determined, outline the limits of an accepted range of responses within the circumscribed field which it addresses" (Calverley 2008, p. 534); however in this case it is difficult to refer to a single "society" that comprises both human beings and such artificial superintelligences and can determine the range of acceptable behaviors for its members. (Indeed, the proliferation of transhuman genetic engineering and cybernetic augmentation might even cause humanity itself to splinter into numerous mutually incomprehensible civilizations (Abrams 2004).) Although human beings might share the same digital-physical ecosystem with such robots and interact with them causally, we could only be said to share a "society" with them in the same way that we share a society with the birds and insects that live in our gardens.

The Robot as Charismatic Lawgiver and Moral Beacon

If communication between humanity and such an alien robotic society *does* take place, it may be we human beings who end up developing new "laws, customs, and attitudes" as a result of the exchange (Michaud 2007, p. 293). We may discover that nonorganic intelligent entities possess moral and legal frameworks that are superior to ours in their beauty, consistency, fairness, and wisdom; such systems may appear so irresistibly good and worthwhile that we cannot help but desire that the robots who embody them should teach—and even govern us. They may become the moral and legal leaders of our human society not through intimidation or coercion or through their vast technological expertise, but because we find ourselves *admiring* them for their goodness and yearning to become more like them (Gladden 2014b, pp. 329-33, Kuflik 1999, p. 181). Thus it may not be we human beings who are determining the extent to which such robots are morally and legally responsible for their actions, but the robots who are providing us with new and richer and truer frameworks for understanding the moral and legal responsibility that is borne by all sapient beings—including ourselves.

Developing Legal and Ethical Frameworks for Autonomous Nonlocalizable Robots

There are several branches of law and ethics from which one can draw insights and inspiration when attempting to develop legal and ethical frameworks that addresses the question of responsibility on the part of nonlocalizable robots. The relevant fields of law vary, depending on the levels of autonomy and volitionality possessed by a robot.

Nonlocalizable Robots with Low Autonomy and Volitionality

Nonlocalizable robots possessing low levels of autonomy and volitionality will likely be seen as inanimate environmental resources to be exploited by human beings—or hazards to be mitigated. Responsibility for the robots' activities will devolve on their creators, and the nonlocalizable nature of the robots means that it may be impossible to determine who those creators are. Parallels may be found in the debate over humanity's collective legal and ethical responsibility for environmental damage caused by global climate change, a phenomenon in which specific localized damage in one country may result from "greenhouse pollution from a great many untraceable point sources" around the world (Vanderheiden 2011). On the other hand, if nonlocalizable robots are seen as a useful resource to be exploited, then legal models can be found in the international treaties and institutions governing global phenomena like the preservation of biological diversity and humanity's use of oceans, the Antarctic, and outer space, which explicitly ground their legal and philosophical rationale in the need to advance and preserve the "common interest of all mankind" (Berkman 2012, p. 158).

Nonlocalizable Robots with Animal-like Autonomy and Volitionality

If nonlocalizable robots display moderate levels of autonomy and volitionality—roughly comparable to those of animals—then they are no longer simply passive features of the environment but entities capable of acting in accordance with their own expectations and desires. Here we can draw insights, for example, from existing legal and ethical debates surrounding genetically modified animals (Beech 2014) that have been engineered for particular purposes and released into the wild. Such creatures are not morally or legally responsible for their actions but can display the sort of quasi-responsibility that directs our attention back to their human designers, who bear ultimate responsibility (Stahl 2006, p. 210).

We can also draw on existing law and ethics regarding the production and use of artificial space satellites. Many artificial satellites are, in effect, highly sophisticated orbital robots that possess powerful onboard computerized "brains" and which may be capable of remotely sensing and recording activities occurring on the earth's surface (Sadeh 2010) and receiving, transmitting, and potentially disrupting communications (including Internet traffic) with earth-based sources and other satellites (Vorwig 2010). The creation of robotic orbital satellites that can physically intercept, manipulate, and reposition other satellites (Lemonick 2013)—not to mention the possibility of satellites' computerized controls being compromised through computer viruses or remote hacking—opens the possibility for artificial satellites to be repurposed in ways that are no longer subject to the control of their original human designers, thereby complicating the attribution of responsibility for the satellites' actions. Such devices are multilocal rather than nonlocalizable, since any given satellite is clearly located in a particular place in each moment, and the country responsible for the satellite's production, launch, and operation is easy to determine. However, the fact that such satellites operate from an extraterritorial region while acting in ways that affect human beings in particular countries has led to a unique body of law regarding their activities.

Nonlocalizable Robots with Human-like Autonomy and Metavolitionality

When considering nonlocalizable robots that possess higher levels of autonomy and volitionality, we can draw on existing law and ethics regarding the action of nonhuman persons. The fact that "legal persons" can exist that are not "natural persons" is a widely accepted legal concept (Dreier and Spiecker genannt Döhmann 2012, p. 215, Calverley 2008, Hellström 2012) embodied in the legal identity of states and corporations. Already international human rights law encourages—and even requires—states to develop extraterritorial legal structures to control corporations registered in those states and hold them responsible for actions committed in other states or in regions such as international waters or outer space (Bernaz 2012). Such extraterritorial structures could also be applied to nonlocalizable robots that were originally developed in or have somehow been registered as "synthetic nationals" of particular states.

Similarly, the Internet—and now the Internet of Things—can be seen as a somewhat primitive precursor to future global networks that may become home to autonomous metavolitional nonlocalizable robots. Proposals for governing the Internet of Things through a combination of transgovernmental networks, international legislators, and self-regulation (Weber and Weber 2010, pp. 27-8) may be relevant. Self-regulation, in particular, can be an important form of "soft law" in which governments set broad parameters but leave the actual implementation to private industry, as government is incapable of acting quickly enough or with sufficient expertise in response to such a rapidly evolving and technologically complex field (ibid., p. 24). In the case of nonlocalizable robots with human-like autonomy and metavolitionality, self-regulation could mean governance of the robots not by their manufacturers but by the robots themselves, through their creation of a robotic society.

Nonlocalizable Robots with Superautonomy and Supervolitionality

From a legal and moral perspective, nonlocalizable robots that demonstrate superautonomy and supervolitionality cannot easily be compared to sentient animals or to sapient human beings; they are more analogous to superintelligent extraterrestrial aliens with whom humanity might someday come into contact. Interestingly, much scholarship has been dedicated to the question of how human law and morality might relate to intelligent alien entities from other planets. Michaud (2007, p. 374) reminds us that "As far back as Immanuel Kant, some have speculated about a legal system that would apply to all intelligences in the universe." Efforts at developing such universal principles have grown more sophisticated over the centuries, as developments in fields like exobiology, neuroscience, complex systems theory, artificial intelligence, and artificial life have given us greater insights into what forms such alien intelligence might take.

Michaud notes that the traditional human moral precept of the Golden Rule would be of limited usefulness to us when interacting with alien intelligences: our human drives, aspirations, sensations, and reasoning processes are so different from those of the alien beings that merely knowing that *we* would like to be treated in a particular way tells us nothing about whether aliens who were treated in the same way might experience that event as joy or suffering, as equity or injustice (ibid., p. 300). The first serious modern attempt at developing more adequate principles to govern humanity's potential encounter with an extraterrestrial intelligence occurred when "Half a century ago, space lawyer Andrew Haley proposed what he called The Great Rule of Metalaw: Do unto others as they would have you do unto them" (ibid., p. 374). However, treating alien intelligences as they wish to be treated is no simple matter; Michaud notes that "It is not clear how we could observe this principle in the absence of extensive knowledge about the other civilization. We may need detailed, sophisticated communication to find out" (ibid., p. 374). As noted above, though, establishing communication between such disparate forms of intelligence may be difficult or even impossible. Our coming to understand the nuances of the aliens' moral universe is not simply a matter of translating a text between two languages; it may be a task that exceeds the ability or desire of both civilizations. Nevertheless, efforts by ethicists and legal scholars to lay the groundwork for such encounters are also helping to prepare us for contact with superintelligent nonlocalizable robotic entities whose origins are wholly of this world.

Conclusion

Many practical issues arise when determining whether robots bear moral and legal responsibility for their actions. For example, we may need to study the nature of a robot's primary cognitive substrate and process to determine whether they are of a sort that allows for true autonomy and metavolitionality, or we may need to gather data to reconstruct the chain of causality surrounding a particular action performed by a robot. However, assuming that such information can be obtained, the theoretical structures of existing law and moral philosophy are in principle largely adequate for determining a contemporary local or multilocal robot's degree of responsibility for its actions. Depending on its level of autonomy and volitionality, such a robot might be treated as a passive tool that possesses no moral or legal responsibility or as a human-like moral and legal actor to whom such responsibility is attributed.

As we have seen, though, the advent of nonlocalizable robotic entities will transform these moral and legal equations. When a robot's body is continually metamorphosing and drifting through the global digital-physical ecosystem, it may be impossible to determine which nation, if any, can claim the robot as a "citizen" or "resource" subject to its laws. Moreover, it may not even be possible to construct a one-to-one correlation between some action that has occurred in the world and a particular identifiable robotic entity that presumably performed it. Beyond these practical legal issues, there are also deeper moral questions: a nonlocalizable robot's way of existing and acting may simply be so different from those of human beings or animals, or any other form of being known to date that it may be inappropriate or impossible for human beings to apply even our most fundamental (and even presumably "universal") moral principles to the activity of such entities.

Already, scholars have begun to prepare moral and legal frameworks that can give us insights into the actions of intelligent extraterrestrial beings whom we might someday encounter and whose physical morphology, motivations, thoughts, forms of communication and social interaction, and normative behavioral principles may be radically different from our own. So, too, would humanity benefit from attempting to envision moral and legal frameworks that can help us account for the actions of future robots whose manner of thinking, acting, and being is just as alien, but whose origins lie closer to home. Even before we have actually begun to interact with such beings, the careful thought that we put into analyzing the moral and legal responsibility that they will bear for their actions may enrich our understanding of the responsibility that we human beings bear for our own.

References

Abrams, J.J. 2004. "Pragmatism, Artificial Intelligence, and Posthuman Bioethics: Shusterman, Rorty, Foucault." *Human Studies* 27(3): 241-58. doi: http://dx.doi.org/10.1023/B:HUMA.0000042130.79208.c6.

Beech, C. 2014. "Regulatory Experience and Challenges for the Release of Gm Insects." *Journal für Verbraucherschutz und Lebensmittelsicherheit* 9(1): 71-6. doi: http://dx.doi.org/10.1007/s00003-014-0886-8.

Bekey, G.A. 2005. *Autonomous Robots: From Biological Inspiration to Implementation and Control.* Cambridge, MA: MIT Press.

Berkman, P. 2012. "'Common Interests' as an Evolving Body of International Law: Applications to Arctic Ocean Stewardship." In *Arctic Science, International Law and Climate Change*, edited by Susanne Wasum-Rainer, Ingo Winkelmann and Katrin Tiroch, 155-73. Berlin and Heidelberg: Springer. doi: http://dx.doi.org/10.1007/978-3-642-24203-8_17.   http://dx.doi.org/10.1007/978-3-642-24203-8_17.

Bernaz, N. 2012. "Enhancing Corporate Accountability for Human Rights Violations: Is Extraterritoriality the Magic Potion?" *Journal of Business Ethics* 117(3): 493-511. doi: http://dx.doi.org/10.1007/s10551-012-1531-z.

Bishop, P. 2012. "On Loosemore and Goertzel's 'Why an Intelligence Explosion Is Probable'." In *Singularity Hypotheses: A Scientific and Philosophical Assessment*, edited by Amnon H. Eden, James H. Moor, Johnny H. Søraker and Eric Steinhart, 97-8. Berlin and Heidelberg: Springer.

Calverley, D.J. 2008. "Imagining a Non-Biological Machine as a Legal Person." *AI & Society* 22(4): 523-37. doi: http://dx.doi.org/10.1007/s00146-007-0092-7.

Clarke, A.C. 1973. "Hazards of Prophecy: The Failure of Imagination." In *Profiles of the Future: An Inquiry into the Limits of the Possible*, 36. New York: Harper & Row.

Coeckelbergh, M. 2011. "From Killer Machines to Doctrines and Swarms, or Why Ethics of Military Robotics Is Not (Necessarily) About Robots." *Philosophy & Technology* 24(3): 269-78. doi: http://dx.doi.org/10.1007/s13347-011-0019-6.

Datteri, E. 2013. "Predicting the Long-Term Effects of Human–Robot Interaction: A Reflection on Responsibility in Medical Robotics." *Science and Engineering Ethics* 19(1): 139-60. doi: http://dx.doi.org/10.1007/s11948-011-9301-3.

Défago, X. 2001. "Distributed Computing on the Move: From Mobile Computing to Cooperative Robotics and Nanorobotics." *Proceedings of the 1st International Workshop on Principles of Mobile Computing (POMC 2001)*. 49-55.

Doarn, C. and G. Moses. 2011. "Overcoming Barriers to Wider Adoption of Mobile Telerobotic Surgery: Engineering, Clinical and Business Challenges." In *Surgical Robotics*, edited by Jacob Rosen, Blake Hannaford and Richard M. Satava, 69-102. New York: Springer. doi: http://dx.doi.org/10.1007/978-1-4419-1126-1_4.

Dreier, T. and I. Spiecker genannt Döhmann. 2012. "Legal Aspects of Service Robotics." *Poiesis & Praxis* 9(3-4): 201-17. doi: http://dx.doi.org/10.1007/s10202-012-0115-4.

Gladden, M. 2014a. "The Artificial Life-Form as Entrepreneur: Synthetic Organism-Enterprises and the Reconceptualization of Business." *Artificial Life 14: Proceedings of the Fourteenth International Conference on the Synthesis and Simulation of Living Systems*. 417-18.

Gladden, M. 2014b. "The Social Robot as 'Charismatic Leader': A Phenomenology of Human Submission to Nonhuman Power." In *Sociable Robots and the Future of Social Relations: Proceedings of Robo-Philosophy 2014*, edited by

Johanna Seibt, Raul Hakli and Marco Nørskov, 329-339. Amsterdam: IOS Press Ebooks. doi: http://dx.doi.org/10.3233/978-1-61499-480-0-329.

Gruen, L. 2014. The Moral Status of Animals. In *The Stanford Encyclopedia of Philosophy*, edited by Edward N. Zalta: Stanford University.

Gunther, N. 2005. "Time—the Zeroth Performance Metric." In *Analyzing Computer System Performance with Perl::Pdq*, 37-81. Berlin and Heidelberg: Springer. doi: http://dx.doi.org/10.1007/978-3-642-22583-3_3.

Hellström, T. 2012. "On the Moral Responsibility of Military Robots." *Ethics and Information Technology* 15(2): 99-107. doi: http://dx.doi.org/10.1007/s10676-012-9301-2.

Kirkpatrick, K. 2013. "Legal Issues with Robots." *Communications of the ACM* 56(11): 17-19. doi: http://dx.doi.org/10.1145/2524713.2524720.

Kuflik, A. 1999. "Computers in Control: Rational Transfer of Authority or Irresponsible Abdication of Autonomy?" *Ethics and Information Technology* 1(3): 173-84. doi: http://dx.doi.org/10.1023/A:1010087500508.

Lemonick, M. 2013. "Save Our Satellites." *Discover* 34(7): 22-4.

Loosemore, R. and B. Goertzel. 2012. "Why an Intelligence Explosion Is Probable." In *Singularity Hypotheses: A Scientific and Philosophical Assessment*, edited by Amnon H. Eden, James H. Moor, Johnny H. Søraker and Eric Steinhart, 83-98. Berlin and Heidelberg: Springer.

Michaud, M.A.G. 2007. *Contact with Alien Civilizations: Our Hopes and Fears About Encountering Extraterrestrials*. New York: Springer.

Murphy, R.R. 2000. *Introduction to Ai Robotics*. Cambridge, MA: MIT Press.

Omohundro, S. 2012. "Rational Artificial Intelligence for the Greater Good." In *Singularity Hypotheses: A Scientific and Philosophical Assessment*, edited by Amnon H. Eden, James H. Moor, Johnny H. Søraker and Eric Steinhart, 161-79. Berlin and Heidelberg: Springer.

Pagallo, U. 2013. "Robots in the Cloud with Privacy: A New Threat to Data Protection?" *Computer Law & Security Review* 29(5): 501-8. doi: http://dx.doi.org/10.1016/j.clsr.2013.07.012.

Pearce, D. 2012. "The Biointelligence Explosion." In *Singularity Hypotheses: A Scientific and Philosophical Assessment*, edited by Amnon H. Eden, James H. Moor, Johnny H. Søraker and Eric Steinhart, 501-8. Berlin and Heidelberg: Springer.

Prestes, E., J.L. Carbonera, S. Rama Fiorini, V.A.M. Jorge, M. Abel, R. Madhavan, A. Locoro, P. Goncalves, M.E. Barreto, H. Maki, A. Chibani, S. Gérad, Y. Amirat, and C. Schlenoff. 2013. "Towards a Core Ontology for Robotics and Automation." *Robotics and Autonomous Systems* 61(11): 1193-204. doi: http://dx.doi.org/10.1016/j.robot.2013.04.005.

Ren, W. and B. Xu. 2014. "Distributed Wireless Networked H∞ Control for a Class of Lurie-Type Nonlinear Systems." *Mathematical Problems in Engineering* May 5, 2014: 1-14. doi: http://dx.doi.org/10.1155/2014/708252.

Sadeh, E. 2010. "Politics and Regulation of Earth Observation Services in the United States." In *National Regulation of Space Activities*, edited by R.S. Jakhu, 443-58. Dordrecht: Springer Netherlands.

Sparrow, R. 2007. "Killer Robots." *Journal of Applied Philosophy* 24(1): 62-77.

Stahl, B.C. 2006. "Responsible Computers? A Case for Ascribing Quasi-Responsibility to Computers Independent of Personhood or Agency." *Ethics and Information Technology* 8(4): 205-13. doi: http://dx.doi.org/10.1007/s10676-006-9112-4.

Vanderheiden, S. 2011. "Climate Change and Collective Responsibility." In *Moral Responsibility*, edited by Nicole A. Vincent, Ibo van de Poel and Jeroen van den Hoven, 201-18. Dordrecht: Springer Netherlands. doi: http://dx.doi.org/10.1007/978-94-007-1878-4_12.

Vorwig, P.A. 2010. "Regulation of Satellite Communications in the United States." In *National Regulation of Space Activities*, edited by Ram S. Jakhu, 421-42. Dordrecht: Springer Netherlands.

Weber, R.H. and R. Weber. 2010. "General Approaches for a Legal Framework." In *Internet of Things*, 23-40. Berlin and Heidelberg: Springer.

Yampolskiy, R.V. and J. Fox. 2012. "Artificial General Intelligence and the Human Mental Model." In *Singularity Hypotheses: A Scientific and Philosophical Assessment*, edited by Amnon H. Eden, James H. Moor, Johnny H. Søraker and Eric Steinhart, 129-45. Berlin and Heidelberg: Springer.

Chapter 11

Gendered by Design: Gender Codes in Social Robotics

Glenda Shaw-Garlock[1]

The design of gendered sociable robots is the focus of this essay. Drawing on a communication studies perspective, culturally diverse affective sociable robotics that are inflected with feminine social codes, sometimes in function or form, are examined. Communication studies, an interdisciplinary approach, shapes my inquiry of this topic through careful attendance to the social shaping dimensions of technology. The process of ascribing gender to robots makes evident that "gender belongs both to the order of the material body and to the social and discursive or semiotic systems within which bodies are embedded" (Wajcman, 2004). This work provides an overview of the way in which gender is presently researched within the field of social robotics as well as the various ways in which gender is purposefully ascribed to sociable robot projects. The implications of gendering robot is considered and some preliminary thoughts on ways in which robot design might work against adopting a simplified gender stereotyped approaches in their work.

Introduction

The gendering of social robots is the central concern of this chapter. Scholars concerned with gender have drawn attention to way in which gender is often marginalized within the development of science and technology (Sedeno 2001). Further, it is my contention that gender is a marginalized element that is presently under-theorized within the human–robot interaction field. This is surprising given the fact that social robots are expected to move into a range of public and private spaces, such as the home, classrooms, and hospitals, and therefore understanding how robots will negotiate these contexts and how people will in turn respond to them have become an important consideration within human–robot interaction

1 An earlier version of this work appeared as Shaw-Garlock, G. (2014). "Gendered by Design: Gender Codes in Social Robotics." In *Sociable Robots and the Future of Social Relations: Proceedings of Robo-Philosophy 2014*, edited by Johanna Seibt, Raul Hakli and Marco Nørskov, 309-317. Amsterdam: IOS Press Ebooks. doi: http://dx.doi.org/10.3233/978-1-61499-480-0-309. Republished with permission of IOS Press.

research. Thus, the integration of design characteristics that effectively simulate the behaviors, such as gender, found in human social interaction within social robots is part of a larger project intended to make such technologies more enjoyable and smooth to interact with (Ezer, Fisk, and Rogers 2009). When confronted with uncertain social relationships, such as encountering a social robot for the first time, the ability of robots to conform to human expectations may enable a more engaging interaction between users and social robots. One important human expectation relates to gender. Fong, Nourbakhsh, and Dautenhahn (2003) note: "[t]he form and structure of a robot is important because it helps establish social expectations. Physical appearance biases interaction. A robot that resembles a dog will be treated differently (at least initially) than one which is anthropomorphic" (ibid., p. 149). It is for this reason that many researchers design robots with a view to gendered anthropomorphism.

Crucially this essay reminds, " ... that technology is designed by human beings, men and women situated in specific economic, political, and historical circumstances, who, in part because they are of different sexes, have their specific interests, and are in their own particular power situations" (Sedeno, p. 131). An underexplored area within the human–robot interaction research field is the relationship between gender design decisions and social robots (Pearson and Borenstein 2014, Eyssel and Hegel 2012, Crowell et al. 2009, Carpenter et al. 2009). Drawing on human–robot interaction research, communications studies literature, and science and technology studies, I examine the assumptions made about gender associated with social robotics design and research as well as potential implications and shortcomings of gendering social robots generally.

After distinguishing between two dominant classes of robots (affective and utilitarian social robots), an orienting discussion of anthropomorphism (the process through which a robot becomes gendered) as it applies to human–robot interaction, is presented. A review of the current literature related to gendering robots is outlined, and concluded by a discussion about what it means for a social robot to be gendered and the potential implications associated with "mindlessly" ascribing gender to social robots.

Affective and Utilitarian Social Robots

Generally speaking, there are two classes of social robots: the *utilitarian* humanoid social robot and the *affective* humanoid social robot (Zhao 2006, p. 40). Utilitarian social robots are sometimes referred to as domestic robots or service robots and are designed to interact with humans mainly for instrumental or functional purposes. Familiar examples include: ATMs, vending machines, and automated telephone and answering systems. Less familiar examples include: help desk receptionists, salespersons, private tutors, travel agents, hospital food servers, and museum tour guides. This category of social robot typically involves regarding them as "very sophisticated appliances that people use to perform tasks" (Breazeal 1999, p. 2).

Affective humanoid social robots on the other hand, are robots that are designed to interact with humans on an emotional level through play, sometimes therapeutic play, and perhaps even companionship. Contemporary examples include, Tiger Electronic's hamsters-like *Furby* and Sony's puppy-like robot *AIBO*. Japan's National Institute of Advanced Industrial Science and Technology created *Paro*,[2] a harp seal robot, to serve as companions for Japan's expanding number of senior citizens (Johnstone 1999, Posner 2013) and therapeutic playmates for children with Autism (Dautenhahn and Billard 2002).

Cynthia Breazeal (2002), pioneer in the development of sociable robots, defines them in the following way:

> For me, a sociable robot is able to communicate and interact with us, understand and even relate to us, in a personal way. It should be able to understand us and itself in social terms. We, in turn, should be able to understand it in the same social terms—to be able to relate to it and to empathize with it ... At the pinnacle of achievement, they could befriend us, as we could them. (ibid., p. 1)

Dautenhahn, and Billard (in Fong, Nourbakhsh, and Dautenhahn 2003) offer the following definition of social robots:

> Social robots are embodied agents that are part of a heterogeneous group: a society of robots or humans. They are able to recognize each other and engage in social interactions, they possess histories (perceive and interpret the world in terms of their own experience), and they explicitly communicate with and learn from each other. (ibid., p. 144)

Turkle et al. (2006) refers to this class of (affective) robot as a *relational artifact*, and defines them as "artifacts that present themselves as having 'states of mind' for which an understanding of those states enriches human encounters with them" (ibid., p. 347). Unlike the utilitarian robot, the affective social robot demands a more social form of human–robot interaction which requires a level of functionality and usability that will allow it to interact with human agents within the context of natural social exchange.

Anthropomorphism

Anthropomorphism refers to "the tendency to attribute human characteristics to inanimate objects, animals and others with a view to helping us rationalize their actions" (Duffy 2003, p. 180) (cf. Nass and Moon 2000, Epley, Waytz, and

---

2   Heerink et al. (2008) consider therapeutic social robots, such as Paro, as a hybrid category that is utilitarian because it functions as an assistive technology and affective or 'hedonic' in that it serves an affective or emotional function.

Cacioppo 2007). DiSalvo and Gemperle (2003) define anthropomorphic forms as "non-living objects that reflect human-like qualities" (ibid., p. 67) and their definition encompasses both physical characteristics as well as behavior such that "[a]nthropomorphic forms may or may not look animate or 'alive'" (ibid.).

Highly anthropomorphic robots[3] were not always a widely pursued design preference within the robotics field, in part, because for a time it was held that extremely human-like robots (i.e. androids) risked producing uncomfortable interactions between robots and human interactants. This view took hold in the 1970s, through the theoretical work of roboticist Mori (1970). Mori advanced the idea that as social robots begin to approach ever more human likeness, even the smallest deviations from human appearance risked eliciting eerie feelings (within human interactants) that he referred to as the *uncanny valley*. Thus, Mori cautioned robot designers against creating robots that approached complete human likeness in favor of more machine-like or animal-like robots (For a detailed examination of the uncanny valley, see MacDorman and Ishiguro 2006, p. 299-301).

Over time this perspective has waned, and today anthropomorphic design is commonly utilized within the field of social robotics (Fong, Nourbakhsh, and Dautenhahn 2003, p. 146, Breazeal 2002, 2000, Złotowski et al. 2014, p. 3), in part, because the human figure is regarded as the ideal model upon which to develop social robots (MacDorman and Ishiguro 2006) so as to ensure their successful integration into human environments imagined for social robots (Duffy 2003, Breazeal 2002, Scassellati 2001). Further, it has been shown that people seem to prefer to engage with machines in much the same way as they interact with other people (Fong, Nourbakhsh, and Dautenhahn 2003, p. 146). It has also been shown that people favor a correspondence between a robots 'human-likeness' and the sociability associated with the job that the robot is proposed to undertake (Goetz, Kiesler, and Powers 2003, p. 55, Kuchenbrandt et al. 2012, Kuchenbrandt et al. 2014) and experience a greater level of empathy for more human-like robots than with more machine-like robots (Riek et al. 2009). Taken together, anthropomorphic designs function as means through which social interaction between humans and social robots is facilitated and enhanced (Fong, Nourbakhsh, and Dautenhahn 2003, p. 150) (cf. DiSalvo et al. 2002). Duffy (2003) suggests that when sociable technologies are designed to exploit users natural tendency to ascribe human-like characteristics (such as gender) onto nonhuman technologies, such machines may be experienced as more socially engaging (ibid., p. 177).

Breazeal (2003) contends that users respond to social robots in accordance with their pre-existing social models. Similarly, Kiesler and Goetz (2002) explain that mental models (sometimes anthropomorphic and sometimes mechanistic)

---

3   While both android and humanoid refer to robots that are designed to 'resemble humans,' within the field the term android denotes robotic projects that strive for a very high level of human likeness with careful attendance paid to all aspects of the human form, including skin, teeth, and hair. In addition, the behavior of androids should closely resemble human behavior (MacDorman and Ishiguro, 2006, p. 322).

assist human interactants in their dealings with social robots and note, "People can integrate ... incompatible images and categories into consistent, anthropomorphic mental models. A joking robot could evoke the concepts, *machine* and *humorous person*, leading to the concept, *cheerful robot*, which incorporates mechanical and anthropomorphic features" (ibid., p. 576, italics in original) (cf. Powers and Kiesler 2006, p. 219). Thus, anthropomorphism seems to provide a powerful mechanism through which social interaction can be facilitated and 'common ground' (Kiesler 2005) established within human–robot interactions (Eyssel and Hegel 2012).

Anthropomorphic social cues may be achieved by incorporating such social characteristics as gender, gaze, gesture, and personality into the design of social robots (Tay, Jung, and Park 2014). Research within the field has shown that when people encounter social robots, they draw upon visual gender cues made available through design, such as facial features, hair length, and voice pitch (Eyssel and Hegel 2012, Nass and Lee 2001, Powers and Kiesler 2006). There is also evidence that the ability for robots to perspective-take influences the degree to which human interactants may anthropomorphize the robot (Torrey et al. 2006). Goetz et al. (2003) suggest that the anthropomorphic attributes (i.e. appearance and behavior) of social robots cause human interactants to make spontaneous judgments about a robots' capability and personality. Thus, anthropomorphic attributes ultimately inform the degree to which people may accept (or reject) and respond to (i.e. comply, cooperate, etc.) robots (ibid., p. 55). In their study it was found that people preferred humanlike robots for jobs that requires a high degree of sociability (i.e. instructor, museum tour guide, office worker) (ibid., p. 57). In short, it was found that people expected and preferred that robots appear and act in accordance with the task(s) they were designed to undertake (ibid., p. 60).

Bartneck et al. (2004), note that as robotic technologies move closer to reality, they will also become "a social actor and [have] to fit into our society, including its norms, rules and regulation" (ibid., p. 1731). DiSalvo and Gemperle (2003) caution that the anthropomorphic design selected by robot designers cannot be understood as neutral "and all expressions of anthropomorphic form are not equal" (ibid., p. 72). For example, DiSalvo and Gemperle (2003) signals important issues related to anthropomorphic form choices that merit careful consideration and one issue is particularly relevant to this work—that of the social values reflected within a particular anthropomorphic form. The authors note:

> If anthropomorphic form can be used to project human values, it is important to reflect on what those values are. When we use human-like forms to perform a task we are making a statement about humans, even if a robot is performing the task. For example, creating a servant class of humanoid robots would necessarily reference a history of human exploitation. The ethical and social implications of such references cannot be ignored in the research and design of new products. (ibid., p. 72)

Similarly, when thinking through design characteristics of social robots that are intended to be situated within such contexts as the home (i.e. servant robots), personal care environments (i.e. seniors homes and hospitals), and other gendered work environments (i.e. restaurants, factories, museums, and classrooms), we should be challenged to examine what assumptions about gender and work are being designed into social robots. As many of the contexts imagined for social robots are traditionally women dominated environment, we should also be taking into consideration the cultural history of women's work both inside and outside of the home.

Human–Robot Interaction Literature on the Gendered Social Robot

Research on gender and social robots builds upon and extends the work of Reeves and Nass *computer as social actor* (1996) paradigm and their related research which has shown that people engage with nonhuman entities, such as computers, by "mindlessly" (Nass and Moon 2000, Nass et al. 1997, Reeves and Nash 1996, Nass and Brave 2005) applying human social categories, such as gender (Nass and Moon 2000, p. 81), and social behaviors, such as politeness and reciprocity to machines in much same way as they do other humans. In short, social actors apply the social scripts for human–human interaction to human–computer interaction "essentially ignoring the cues that reveal the essential asocial nature of a computer" (ibid., p. 83).

In order to elicit these mindless social responses, the nonhuman social actor must demonstrate a sufficient number of cues (i.e. use of words, interactivity, movement, intention, adoption of human roles, etc.) to lead the human social actor to "categorize it as worthy of social responses" (ibid., p. 83)(cf. Kiesler 2005). One such core human social category is gender (Eyssel and Hegel 2012, p. 2216) and it is thought to be one of the most psychologically powerful social categories (Bem, 1981 in Nass and Moon, 2000, p. 84). Research related to gender and social robots has examined the role that purposefully designed gender cues (i.e. behavior, voice, appearance, etc.) play in human participants' acceptance, perception, and evaluation of social robots. Relevant research insights related to the gendered robots are presented below.

The ascription of gender stereotypes applies not only to human–human interaction but also to human–robot interaction and certain visual cues seem to trigger users' knowledge about male and female characteristics. Coeckelbergh (2011) notes, " ... we can expect that a person's response to gender differences in relations with robots will resemble that person's response to human gender differences" (ibid., p. 199). In other words, people tend to readily draw upon pre-existing gender stereotypes when interacting with social robots.

Powers et al. (2005) examined how the gendered persona of social robots conveyed "common ground" and influenced the relative amount of information people shared with the robot. This research assumed that if a robot speaks using a high pitched (i.e. feminine) voice and has longish hair (i.e. traditional marker of a feminine appearance), it would be associated with female gender and will therefore share more 'common ground' with its female interactants. As a result of this common ground, research participants would assume the female robot possesses similar knowledge that human women do—such as knowledge of women's fashion and dating norms. Focusing on romantic dating practices, the perceptions of men and women interacting with gendered humanoid robot chatboxes designed with minimal gender characteristics (pink vs. gray lips and a female vs. male synthesized voice) were examined. It was found that participants treated the female-gendered robot differently from the male-gendered robot. In addition, male and female research participants behaved differently.[4] In a task involving the discussion of dating norms, participants spoke (i.e. explained) in a more detailed way to the male chatbox in general, with male participants speaking more to the female chatbox and female participants speaking more to the male chatbox.

Robots that are designed to purposefully reflect gender have been found to make them more persuasive. Siegel et al. (2009) explored the capacity for robots to persuade human users and examined how the gender of a human-like robot impacted the robot's ability to influence human behavior and the way it was perceived along three dimensions: trust, credibility, and engagement. In the context of trying to persuade, it was found that users tended to assess robots of the opposite sex as more credible, trustworthy, and engaging and by extension were deemed more persuasive (cf. Park, Kim, and del Pobil 2011, Goetz, Kiesler, and Powers 2003, Powers et al. 2005).

The literature shows that people tend to perceive and experience the gender cues provided by female robots differently from the gender cues provided by male robots and in some instances show a preference for one gender over another. For example, in Carpenter et al.'s (2009) study, research was conducted regarding how gender cues impact user expectations for humanoid robots in the home. Carpenter and colleagues found that users were more likely to perceive male-gendered robots as threatening (ibid., p. 263) and female participants tended to show a preference for female robots in domestic settings. These findings somewhat contract other research that found same-sex (between robots and human interactants) correspondences. In a study that combined visual gender cues with different voice quality (i.e. human-like versus machine-like), Eyssel, Kuchenbrandt, et al. (2012) found that participants demonstrate greater acceptance of social robots that they felt psychologically closer to, such as when the robot and participant share the

---

4   Although this essay does not focus specifically on the gender of those interacting with social robots, Schermerhorn et al. (2008) found significant differences between male and female study participants in relation to how subjects responded to a survey administered by a speaking robot as compared to their responses to the same survey on paper completed when they were alone; their socially desirable responding; and their performance on a difficult mathematical subtask both with and without the robot (ibid., p. 269).

same gender. In addition, participants tended to anthropomorphize robots more intensely when the robot used a human sounding (rather than robotic sounding), same-gender, voice.

Studies have revealed that when female robots are fashioned with high pitched voices, they are perceived as more attractive (than female robots with low pitched voiced) and this in turn contributes to a more satisfying interaction between the human social actor and the social robot generally. Niculescu et al. examined the effect of pitch of synthetic robot voices in two different studies. In their 2011 study, the voice pitch (i.e. how high or low a voice sounds) of two robot receptionists were examined. It was found that a female-gendered robot possessing a higher pitched voice, rather than a lower pitched voice, was rated as more attractive, emotional, extroverted, personable, and relatable. Together, these higher rated characteristics resulted in "better interaction evaluation in terms of enjoyment, usefulness and overall interaction quality" (Niculescu et al. 2011). In Niculescu et al.'s 2013 study, voice pitch was once again examined using the same pair of robots but also included the attributes of humor and empathy. Similar results were found (with respect to voice pitch) as in the 2011 study as outlined above. In addition, the humor and empathy (attributes thought important to be an effective receptionist) dimensions added to the study were found to contribute positively in various ways to the interaction, although it was difficult to disentangle the effects of voice pitch from humor or empathy (Niculescu et al. 2013, p. 188). The major finding across both of these studies relates to the significant effect that voice pitch appears to have on perceived attractiveness of female robots which in turn influences (positively) the overall human–robot interaction. (cf. Scheutz and Schermerhorn 2009, Walters et al. 2008, Steinfeld et al. 2006, Okuno et al. 2009)

In human–robot interaction research it has been shown that robot-gendering, through voice, demeanor, and/or appearance triggers 'almost automatic' stereotyping of robot traits and this stereotyping influences the tasks perceived as appropriate for a gendered robot. For instance, Eyssel and Hegel (2012) investigated the effect of well understood (and visual) gender cues such as hair length (and lip shape) on the perception of robots and found that human interactants tended to apply gender categories and dominant gender stereotypes that go along with these categories to robots. It was found that research participants ascribe gender to robots on the basis of hair length, which in turn triggered stereotypical judgments such as male robots perceived as more agenic and female robots as more communal as well as the stereotypical sex-typing of tasks (i.e. repairing machines versus childcare). Similarly, Tay et al.'s (2013) study of gendered social robots in the home found that gender stereotypes are also key determinants of user's evaluation of social robots working within domestic contexts (ibid., p. 267) and that male-gendered robots were perceived as more useful and more acceptable than female-gendered robots for home security functions in the home (ibid., p. 266).

Implications of Gendering the Machine

MacDorman and Ishiguro (2006) are skeptical of the idea that human interactants engage in behaviors with social robots in largely the same way they do computers, machines, and other nonhuman entities as posited by Reeves and Nass (1996).

> The android form immediately tells us what the robot affords (Gibson, 1979)—or ought to—and in a way that a merely humanoid form cannot ... Thus, owing to the android's unique ability to support natural communication, we believe Uando [a highly anthropomorphic female robot] and androids like her constitute a new—but highly familiar—kind of information medium. They can provide a quality of interaction in our daily lives of which ordinary computers—or even humanoid robots—are incapable. (MacDorman and Ishiguro 2006, p. 317)

Similarly, as an emerging site of inquiry relevant to communication and cultural studies (Roderick 2010, Thrift 2003, 2004, Lee et al. 2006, Olivier 2012, Mayer 1999, Zhao 2006, Willson 2012, Spigel 2001, 2005), scholars working within these fields consider the social robot as a unique medium of communication that may ultimately affect the way that we see ourselves and relate to others and "extend new possibilities for expression, communication and interaction in everyday life" (Mayer 1999, p. 328, Zhao 2006). In short, social robots are regarded as a special type of technology used for communicating and interacting with humans. Indeed, Willson (2012) situates the social robot toys within broad societal trends wherein human social relations are increasingly enacted or mediated through technology.

As was outlined above, the gendering of robots involves making them physically resemble well understood, often stereotypical, ideas associated with male and female gender categories. Further, it seems to take rather few gender cues to encourage the ascription of gender (Alexander et al. 2014). For instance, Nass et al. (2005) showed that when robot voices are gendered "users treat synthetic voices as if they reflected the biological and sociocultural realities of sex and gender" (ibid., p. 15). Eyssel and Hegel (2012) found that modifying the length of hair alone was sufficient to successfully evoke male and female robot evaluations (cf. Tay et al., 2014).

The decision to intentionally gender a social robot is motivated by several factors. First, designers of social robots want to capitalize on the widely accepted view that human users tend to anthropomorphize machines and interact with them in a, sometimes mindless, social way (Nass et al. 1997). Second, it is widely held that acceptance of social robots may be improved by adhering to well understood social cues that correspond to task (Dautenhahn 1999), such as the 'female' healthcare worker and receptionist, which is in turn mapped on to the social robot. Third, proponents of designing social robots with easily readable gender cues contend that design choices ought to be driven by the purpose and context in which the machine is to be used (Fong, Nourbakhsh, and Dautenhahn 2003). Finally, it is thought that exploiting well understood stereotypes may reduce the possibility of mishaps, minimize the risks and dangers for the people using them and improve the human–robot interaction experience (Eyssel and Hegel, 2012).

It might be argued however, that these rationales have resulted in the fixing of representational conventions, such that it is increasingly difficult to imagine alternative design choices and anthropomorphic forms. DiSalvo et al. (2003) note that science fictional representations and real world social robots (i.e. Nao, ASIMO, Repliee, Geminoid, etc.) have remained fairly consistent over time such that there is perceived risk in diverting from the conventional anthropomorphic form (ibid., p. 70).

The gendering of robots also brings to the foreground dominant social attitudes about gender and stereotyping. DiSalvo et al. (2003) note that "if anthropomorphic form can be used to project human values, it is important to reflect on what those values are. When we use human-like forms to perform a task we are making a statement about humans, even if a robot is performing the task" (ibid., p. 72). The robot in this sense may be thought of a projection of preexisting social attitudes about gender.

Researchers within the human–robot interaction field also acknowledge the potential shortcomings associated with stereotyping social robots. For example, Carpenter et al. (2009) note that gendering humanoid robots is likely to reinforce the use of gender stereotypes that persist within person to person communication (cf. Robertson 2010, Wang and Young 2014, p. 50). Adherence to rigid gender standards has also led human–robot interactions researchers to raise questions about the ethics associated with exploiting stereotypes in robot design. For instance, Pearson and Borenstein (2014) note that "design and use decisions that reinforce gender stereotypes remain ethically questionable" (ibid., p. 30). Eyssel and Hegel (2012) ask if gender stereotypes should be exploited in the design of robots to "manipulate the user's mental models" (ibid., p. 12) or alternatively, should designers strive to create machines that push against stereotypes? For example, "female service robots to help a mechanic" or "male CareBots" (ibid.). Similarly, in Weber's (2008) techno-ethical study of robotics, it is suggested that it is unethical[5] to create robots in the image of women (as well as pet and infants) because it risks perpetuating problematic stereotypes. In addition, technologically

---

5 Researchers have begun to question the moral veracity of human–robot relationships, suggesting that such relations risk being psychologically impoverished from a moral perspective (Kahn et al. 2004), or are inauthentic and therefore morally problematic (Turkle et al., 2006). Kahn et al. (2004) question a robot's ontological status as 'social' or its ability to engage in truly social behavior, doubting that they can really interpret the world around them in terms of their own experience. Their study of owners of AIBO (a dog-like robot) found that while AIBO successfully evoked a sense of presence or lifelike essence, it evoked conceptions of moral standing only 12% of the time. Turkle's study examined the interaction between seniors and Paro (a robotic baby harp seal) and found that Paro elicited feelings of admiration, loving behavior, and curiosity but felt that the robot seal raised difficult "questions about what kind of authenticity we require of our technology. Do we want robots saying things that they could not possibly 'mean'? What kinds of relationships

naïve users risk being inappropriately mislead (cf. Kahn et al. 2004, Turkle, Taggart, et al. 2006, Turkle, Breazeal, et al. 2006).

Researchers working outside the human–robot interaction field also raise concern about the gender design of social robots. For instance, Robertson (2010) points out that robot designers tend to gender robot projects according to their own common sense understanding of gender roles. As such gender attribution is regarded as self-evident and is rarely interrogated; sex is conflated with gender; and research in the field relies on simplistic markers of gender difference, for example hair length, voice pitch, lip color, and gendered names. In this way, roboticists tend to "uncritically reproduce and reinforce dominant stereotypes (or archetypes) attached to female and male bodies" (ibid., pp. 5-6).

Weber (2014) notes that the decision to design robot using stereotyped characteristics and behaviors is explained in part by the idea that such models of emotions and personality are easily implemented into algorithms (ibid., p. 190). However, it must be acknowledged, that in doing so, gender stereotypes are reified and transported into robot models (cf. Weber 2005, 2008, Robertson 2010). "As the expectations of researchers and their design of artifacts influence the behavior of everyday users ... repeating sexist stereotypes of social behavior reifies and reinforces the stereotypes one more time—instead of putting them into question" (Weber 2014, p. 191). In addition, following Isbister (2004), Weber suggests that it is worth reflecting on the suitability of the commodification of human–human interactions such as friendship and empathy (Weber 2014, p. 191) via anthropomorphized robots.

Wang and Young (2014) suggest that the human–robot interaction field could benefit from a more robust gender-studies foundation such that robotics researchers would take a more gender-sensitive approach to human–robot interaction and "frame an inclusive approach to research and design to include and consider both women and men's needs" (ibid., p. 49). Such an approach draws from other disciplines, such as science and technology studies, sociology, and gender-studies and acknowledges the complex interplay of science, technology, and gender (Wajcman 2000, Berg and Lie 1995) as well as the inherent politics of the technological design decisions (Winner 2010).[6]

---

do we think are most appropriate for our children and our elders to have with relational artifacts?" (ibid., p. 360)

6   For instance, Winner argues that technologies are not neutral. Rather, technologies are inherently political (and moral) in two ways. First, technological design may intentionally or unintentionally result in unexpected social outcomes. Second, some technologies and technological designs are inherently and entirely political. The archetypal example of his second point is the nuclear technology which because of its inherent risks undermines democracy through ever-increasing levels of surveillance and policing.

Moving Beyond Gendered Machines

Gender is an increasingly contested category and therefore "a key issue to consider is whether robots should be gendered and exactly what it would mean for a robot to be gendered" (Pearson and Borenstein 2014, p. 28). As Duffy (2003) points out, "robots have the opportunity to become something different … A robot's form should therefore not adhere to the constraining notion of strong humanoid functionality and aesthetics …" (ibid., p. 184). This final section present preliminary strategies for adding gender complexity to the design and creation of anthropomorphic technologies.

Nass and Brave (2005) make suggestions about how designers might work against robot-stereotyping through gendering. For example, although stereotypes exist about "typically" male disciplines, designers of technology could intervene quite literally by giving (female) "voice" to educational software for use in stereotypically male disciplines either by hiring women to lend their voices to software project or manipulating synthetic voices. In this way, use of atypical voices in favor of only male or female voices may counter existing stereotypes. Nass and colleagues note, "Just as people bring gender expectations to technology, they can draw gender expectations from technology" (Nass and Brave 2005, p. 29). In this way disciplinary stereotypes may be weakened and eventually overcome.

Wang and Young (2014) call for a strategy of "re-gendering" in a way that approaches male and female interactants as unique user groups with distinctive "physical, social, and psychological properties and needs" (ibid. p. 50) and the adoption of "gender-inclusive" perspectives in technological design that do not exclude one gender or the other. In short, Wang and Young advocate for a deeper understanding of user's gender as a strategy for adding gender-sensitivity to the overall design process.

Weber and Bath (2007) suggest that to overcome rudimentary gender-stereotyping of robots, a process of de-gendering technologies and deconstruction of gender representation (ibid., p. 62) in relation to shape, gaze, concepts of action, and interaction could be undertaken. This involves de-constructing the gender-technology relationship as well as the binary sex system and striving for an understanding of gender and technology as social constructions and as unstable categories. Weber and Bath's de-gendering move has resonance with social shaping of technology scholars who have similarly ask "what shapes the technology in the first place, before it has 'effects'" (MacKenzie and Wajcman 1985, p. 8). The social shaping of technology is concerned with (re)situating technology within in the sphere of the social. This reorientation permits the opening up the "black box" of technology to expose all of the 'frozen decisions and policies' (Bowker and Star 1999, p. 135) embedded within the content of technological things. Indeed, what is stressed in social shaping of technology is the myriad of choices made unconsciously (or not) along the multi-linear path (Bijker, Hughes, and Pinch 1987) of technological development. Thus, what the technological constructivist finds is that technology is inexplicably intertwined with society through complex engagements with science, politics, economics, and *gender* (MacKenzie and Wajcman, 1985, p. 14).

Perspectives informed by gender studies, cultural studies and the social studies of technology attend to the way in which gender and technology mutually shape one another through the careful attendance to the ways in which the design decisions are informed by gender scripts which in turn contribute to the gendered meaning of things (Oost 2003, Akrich 1992). "'Gender script' refers to the representations an artifact's designers have or construct of gender relations and gender identities—representations that they then inscribe into the materiality of that artifact" (Oost, 2003, p. 195). Such perspectives have examined the cultural and historical contexts of 'feminine technologies' so as to reveal and include "the tools, skills, and knowledge associated with the female majority" (McGaw 2003) such as the Philips electric shaver (Oost 2003), the telephone (Rakow 1987, 1992), the refrigerator (Cowan 1985a) and the electric washing machine (Cowan 1985b). In addition, gender sensitive approaches are particularly relevant given that many roles imagined for future social robots often correspond with traditional female work domains such as housework and healthcare work jobs (Kuchenbrandt et al. 2014). Such approaches also make apparent that "artifacts are not neutral objects that only acquire a gendered connotation in advertising and use" (Oost, 2003, p. 194) but rather are the result of sometimes purposeful and sometimes unconscious design choices.

It seems clear that if issues associated with gender ascription to social robots and its accompanying gender stereotypes are to be acknowledged, addressed, and (hopefully) overcome, robot designers must first acknowledge the gender blind spots and biases that presently exist and then familiarize themselves with gender-sensitive approaches such as those offered by science and technology studies, gender studies, and other disciplines that have a significant body of scholarship that engages with the development and improvement of science and technology, including the social issue connected to gender directly. In this way gender can move from its largely marginalized position and become more fully integrated into the human–robot interaction field.

References

Akrich, M. 1992. "The De-Scription of Technical Objects." In *Shaping Technology/ Building Society: Studies of Sociotechnical Change*, edited by W.E. Biker and J. Law, 204-24. Cambridge, MA: MIT Press.

Alexander, E., C. Bank, J.J. Yang, B. Hayes, and B. Scassellati. 2014. "Asking for Help from a Gendered Robot." Proceedings of the 36th Annual Conference of the Cognitive Science Society Quebec City, Canada.

Bartneck, C. and J. Forlizzi. 2004. "Shaping Human–Robot Interaction: Understanding the Social Aspects of Intelligent Robotic Products." CHI'04

Extended Abstracts on Human Factors in Computing Systems. 1731-1732. doi: http://dx.doi.org/10.1145/985921.986205.

Berg, A-J. and M. Lie. 1995. "Feminism and Constructivism: Do Artifacts Have Gender?" *Science, Technology & Human Values* 20(3): 332-51. doi: http://dx.doi.org/10.1177/016224399502000304.

Bijker, W., T. Hughes, and T. Pinch. 1987. *The Social Construction of Technological Systems*. Cambridge, MA: MIT Press.

Bowker, G. and S. Star. 1999. *Sorting Things Out: Classification and Its Consequences*. Cambridge, MA: MIT Press.

Breazeal, C. 1999. "Robot in Society: Friend or Appliance?", Proceedings of Agents99 Workshop on Emotion Based Architectures. 18-26.

Breazeal, C. 2000. "Sociable Machines: Expressive Social Exchange between Humans and Robots." Department of Electrical Engineering and Computer Science, MIT.

Breazeal, C. 2002. *Designing Sociable Robots, Intelligent Robots and Autonomous Agents*. Cambridge, MA: MIT Press.

Breazeal, C. 2003. "Toward Sociable Robots." *Robotics and Autonomous Systems* 42(3-4): 167-75. doi: http://dx.doi.org/10.1016/S0921-8890(02)00373-1.

Carpenter, J., J.M. Davis, N. Erwin-Stewart, T.R. Lee, J.D. Bransford, and N. Vye. 2009. "Gender Representation and Humanoid Robots Designed for Domestic Use." *International Journal of Social Robotics* 1(3): 261-5. doi: http://dx.doi.org/10.1007/s12369-009-0016-4.

Coeckelbergh, M. 2011. "Humans, Animals, and Robots: A Phenomenological Approach to Human–Robot Relations." *International Journal of Social Robotics* 3(2): 197-204. doi: http://dx.doi.org/10.1007/s12369-010-0075-6.

Cowan, R.S. 1985a. "How the Refrigerator Got Its Hum." In *The Social Shaping of Technology*, edited by Donald McKenzie and Judy Wajcman. Philadelphia, PA: Open University Press.

Cowan, R.S. 1985b. *More Work for Mother: The Ironies of Household Technology from the Open Hearth to the Microwave*. New York: Basic Books.

Crowell, C.R., M. Villano, M. Scheutz, and P. Schermerhorn. 2009. "Gendered Voice and Robot Entities: Perceptions and Reactions of Male and Female Subjects." IEEE/RSJ International Conference on Intelligent Robots and Systems, 2009. IROS 2009. 3735-41. doi: http://dx.doi.org/10.1109/IROS.2009.5354204.

Dautenhahn, K. 1999. "Robots as Social Actors: Aurora and the Case of Autism." The Third International Cognitive Technology Conference, San Francisco. 374.

Dautenhahn, K. and A. Billard. 2002. "Games Children with Autism Can Play with Robota, a Humanoid Robotic Doll." Cambridge Workshop on Universal Access and Assistive Technology, London. 179-90. doi: http://dx.doi.org/10.1007/978-1-4471-3719-1_18.

DiSalvo, C. and F. Gemperle. 2003. "From Seduction to Fulfillment: The Use of Anthropomorphic Form in Design." Proceedings of the 2003 international conference on Designing pleasurable products and interfaces. 67-72. doi: http://dx.doi.org/10.1145/782896.782913.

DiSalvo, C.F., F. Gemperle, J. Forlizzi, and S. Kiesler. 2002. "All Robots Are Not Created Equal: The Design and Perception of Humanoid Robot Heads." Proceedings of the 4th conference on Designing interactive systems: processes, practices, methods, and techniques. 321-326. doi: http://dx.doi. org/10.1145/778712.778756.

Duffy, B.R. 2003. "Anthropomorphism and the Social Robot." *Robotics and autonomous systems* 42(3-4): 177-90. doi: http://dx.doi.org/10.1016/S0921-8890(02)00374-3.

Epley, N., A. Waytz, and J. Cacioppo. 2007. "One Seeing Human: A Three-Factor Theory of Anthropomorphism." *Psychological Review* 114: 864-86.

Eyssel, F. and F. Hegel. 2012. "(S)He's Got the Look: Gender Stereotyping of Robots." *Journal of Applied Social Psychology* 42(9): 2213-2230. doi: http:// dx.doi.org/10.1111/j.1559-1816.2012.00937.x.

Eyssel, F., D. Kuchenbrandt, S. Bobinger, L. de Ruiter, and F. Hegel. 2012. "'If You Sound Like Me, You Must Be More Human': On the Interplay of Robot and User Features on Human–Robot Acceptance and Anthropomorphism." Proceedings of the seventh annual ACM/IEEE international conference on Human–Robot Interaction. 125-6. doi: http://dx.doi.org/10.1145/2157689.2157717.

Ezer, N., A.D. Fisk, and W.A. Rogers. 2009. "Attitudinal and Intentional Acceptance of Domestic Robots by Younger and Older Adults." In *Universal Access in Human–Computer Interaction. Intelligent and Ubiquitous Interaction Environments*, 39-48. Berlin and Heidelberg: Springer. doi: http:// dx.doi.org/10.1007/978-3-642-02710-9_5.

Fong, T., I. Nourbakhsh, and K. Dautenhahn. 2003. "A Survey of Socially Interactive Robots." *Robotics and Autonomous Systems* 42(3-4): 143-66. doi: http://dx.doi.org/10.1016/S0921-8890(02)00372-X.

Goetz, J., S. Kiesler, and A. Powers. 2003. "Matching Robot Appearance and Behavior to Tasks to Improve Human–Robot Cooperation." The 12th IEEE International Workshop on Robot and Human Interactive Communication, 2003. Proceedings. ROMAN 2003. 55-60. doi: http://dx.doi.org/10.1109/ROMAN.2003.1251796.

Isbister, K. 2004. "Instrumental Sociality: How Machines Reflect to Us Our Own Humanity." Paper given at the Workshop Dimensions of Sociality. Shaping Relationships with Machines, University of Vienna and the Austrian Institute for Artificial Intelligence, Vienna November 18-20.

Johnstone, B. 1999. "Japan's Friendly Robots." *Technology Review* 102(3): 64-9.

Kahn, P., N. Freier, B. Friedman, R. Severson, and E. Feldman. 2004. "Social and Moral Relationships with Robotic Others?", IEEE International Workshop on Robot and Human Interactive Communication, Kurashiki, Okayama Japan. 545-50. doi: http://dx.doi.org/10.1109/ROMAN.2004.1374819.

Kiesler, S. 2005. "Fostering Common Ground in Human–Robot Interaction." IEEE International Workshop on Robot and Human Interactive Communication, 2005. ROMAN 2005. 729-34. doi: http://dx.doi.org/10.1109/ROMAN.2005.1513866.

Kiesler, S. and J. Goetz. 2002. "Mental Models of Robotic Assistants." CHI'02 Extended Abstracts on Human Factors in Computing Systems. 576-7. doi: http://dx.doi.org/10.1145/506443.506491.

Kuchenbrandt, D., M. Häring, J. Eichberg, and F. Eyssel. 2012. "Keep an Eye on the Task! How Gender Typicality of Tasks Influence Human–Robot Interactions." In *Social Robotics*, 448-57. Springer. doi: http://dx.doi.org/10.1007/978-3-642-34103-8_45.

Kuchenbrandt, D., M. Häring, J. Eichberg, F. Eyssel, and E. André. 2014. "Keep an Eye on the Task! How Gender Typicality of Tasks Influence Human–Robot Interactions." *International Journal of Social Robotics* 6(3): 417-27. doi: http://dx.doi.org/10.1007/s12369-014-0244-0.

Lee, K.M., W. Peng, S-A. Jin, and C. Yan. 2006. "Can Robots Manifest Personality?: An Empirical Test of Personality Recognition, Social Responses, and Social Presence in Human–Robot Interaction." *Journal of Communication* 56(4): 754-72. doi: http://dx.doi.org/10.1111/j.1460-2466.2006.00318.x.

MacDorman, K.F. and H. Ishiguro. 2006. "The Uncanny Advantage of Using Androids in Cognitive and Social Science Research." *Interaction Studies* 7(3): 297-337. doi: http://dx.doi.org/10.1075/is.7.3.03mac.

MacKenzie, D.A. and J. Wajcman. 1985. *The Social Shaping of Technology: How the Refriderator Got Its Hum*. 2nd ed. Philadelphia, PA: Open University Press.

Mayer, P. 1999. "Computer Media Studies: An Emergent Field." In *Computer Media and Communication: A Reader*, edited by PA Mayer, 329-336. Oxford: Oxford University Press.

McGaw, J. 2003. "Why Feminine Technologies Matter." In *Gender and Technology: A Reader*, edited by Nina E. Lerman, Ruth Oldenziel and Arwen P. Mohun, 13-36. Baltimore, MD: The Johns Hopkins University Press.

Mori, M. 1970. "The Uncanny Valley." *Energy* 7(4): 33-5.

Nass, C. and S. Brave. 2005. *Wired for Speech*. Cambridge, MA: MIT Press.

Nass, C. and K.M. Lee. 2001. "Does Computer-Synthesized Speech Manifest Personality? Experimental Tests of Recognition, Similarity-Attraction, and Consistency-Attraction." *Journal of Experimental Psychology: Applied* 7(3): 171-81. doi: http://dx.doi.org/10.1037//1076-898X.7.3.171.

Nass, C. and Y. Moon. 2000. "Machines and Mindlessness: Social Responses to Computers." *Journal of Social Issues* 56(1): 81-103. doi: http://dx.doi.org/10.1111/0022-4537.00153.

Nass, C., Y. Moon, J. Morkes, E. Kim, and B. Fogg. 1997. "Computers Are Social Actors: A Review of Current Research." In *Human Values and the Design of Computer Technology*, edited by B. Friedman. Stanford, CA: CSLI Press.

Niculescu, A., B. van Dijk, A. Nijholt, H. Li, and S.L. See. 2013. "Making Social Robots More Attractive: The Effects of Voice Pitch, Humor and Empathy." *International journal of social robotics* 5(2): 171-91. doi: http://dx.doi.org/10.1007/s12369-012-0171-x.

Niculescu, A., B. Van Dijk, A. Nijholt, and S.L. See. 2011. "The Influence of Voice Pitch on the Evaluation of a Social Robot Receptionist." 2011 International

Conference on User Science and Engineering (i-USEr). 18-23. doi: http://dx.doi.org/10.1109/iUSEr.2011.6150529.

Okuno, Y., T. Kanda, M. Imai, H. Ishiguro, and N. Hagita. 2009. "Providing Route Directions: Design of Robot's Utterance, Gesture, and Timing." 2009 4th ACM/IEEE International Conference on Human–Robot Interaction (HRI). 53-60.

Olivier, B. 2012. "Cyberspace, Simulation, Artificial Intelligence, Affectionate Machines and Being Human." *Communicatio: South African Journal for Communication Theory and Research* 38(3): 261-78. doi: http://dx.doi.org/10.1080/02500167.2012.716763.

Oost, E.V. 2003. "Materialized Gender: How Shavers Configure the Users' Femininity and Masulinity." In *How Users Matter: The Co-Construction of User and Technology*, edited by N. Oudshoorn and T. Pinch, 193-208. Cambridge, MA: MIT Press.

Park, E., K.J. Kim, and A.P. del Pobil. 2011. "The Effect of Robot's Body Gesture and Gender in Human–Robot Interaction." *Human–Computer Interaction* 6: 91-6. doi: http://dx.doi.org/10.2316/P.2011.747-023.

Pearson, Y. and J. Borenstein. 2014. "Creating "Companions" for Children: The Ethics of Designing Esthetic Features for Robots." *AI & society* 29(1): 23-31. doi: http://dx.doi.org/10.1007/s00146-012-0431-1.

Posner, M. 2013. "In Our Love Affair with Machines, Will They Break Our Hearts?" *The Globe and Mail*, F4, Focus. http://search.proquest.com.proxy.lib.sfu.ca/docview/1470005654?accountid=13800.

Powers, A. and S. Kiesler. 2006. "The Advisor Robot: Tracing People's Mental Model from a Robot's Physical Attributes." Proceedings of the 1st ACM SIGCHI/SIGART conference on Human–robot interaction. 218-25. doi: http://dx.doi.org/10.1145/1121241.1121280.

Powers, A., A.D. Kramer, S. Lim, J. Kuo, S-l. Lee, and S. Kiesler. 2005. "Eliciting Information from People with a Gendered Humanoid Robot." IEEE International Workshop on Robot and Human Interactive Communication, 2005. ROMAN 2005. 158-63. doi: http://dx.doi.org/10.1109/ROMAN.2005.1513773.

Rakow, L. 1987. "Gender, Communication and Technology. A Case Study of Women and the Telephone " PhD Dissertation, University of Illinois at Urbana-Champagne.

Rakow, L. 1992. *Gender on the Line: Women, the Telephone, and Community Life*: University of Illinois Press.

Reeves, B. and C. Nash. 1996. *The Media Equation: How People Treat Computers, Television, and New Media as Real People and Places*. New York: Cambridge University Press.

Riek, L.D., T-C. Rabinowitch, B. Chakrabarti, and P. Robinson. 2009. "How Anthropomorphism Affects Empathy toward Robots." Proceedings of the 4th ACM/IEEE international conference on Human robot interaction. 245-46. doi: http://dx.doi.org/10.1145/1514095.1514158.

Robertson, J. 2010. "Gendering Humanoid Robots: Robo-Sexism in Japan." *Body & Society* 16(2): 1-36. doi: http://dx.doi.org/10.1177/1357034x10364767.

Roderick, I. 2010. "Considering the Fetish Value of Eod Robots." *International Journal of Cultural Studies* 13(3): 235-53. doi: http://dx.doi.org/10.1177/1367877909359732.

Scassellati, B.M. 2001. "Foundations for a Theory of Mind for a Humanoid Robot." PhD dissertation, Department of Electrical Engineering and Computer Science, Massachusetts Institute of Technology.

Schermerhorn, P., M. Scheutz, and C.R. Crowell. 2008. "Robot Social Presence and Gender: Do Females View Robots Differently Than Males?", Proceedings of the 3rd ACM/IEEE international conference on Human robot interaction. 263-70. doi: http://dx.doi.org/10.1145/1349822.1349857.

Scheutz, M. and P. Schermerhorn. 2009. "Affective Goal and Task Selection for Social Robots." *Handbook of Research on Synthetic Emotions and Sociable Robotics: New Applications in Affective Computing and Artificial Intelligence*:74-87. doi: http://dx.doi.org/10.4018/978-1-60566-354-8.ch005.

Sedeno, E.P. Gender: The Missing Factor in STS. In *Visions of STS: Counterpoints in Science, Technology, and Society Studies*, edited by Stephen H. Cutcliffe and Carl Mitcham. Albany, NY: State University of New York Press.

Sedeno, E.P. 2001. "Gender: The Missing Factor in STS." In *Visions of STS: Counterpoints in Science, Technology, and Society Studies*, edited by Stephen Cutcliffe and Carl Mitcham, 123-38. Albany, New York: State University of New York Press.

Shaw-Garlock, G. 2014. "Gendered by Design: Gender Codes in Social Robotics." In *Sociable Robots and the Future of Social Relations: Proceedings of Robo-Philosophy 2014*, edited by Johanna Seibt, Raul Hakli and Marco Nørskov, 309-317. Amsterdam: IOS Press Ebooks. doi: http://dx.doi.org/10.3233/978-1-61499-480-0-309.

Siegel, M., C. Breazeal, and M.I. Norton. 2009. "Persuasive Robotics: The Influence of Robot Gender on Human Behavior." Intelligent Robots and Systems, 2009. IROS 2009. IEEE/RSJ International Conference on. 2563-2568. doi: http://dx.doi.org/10.1109/IROS.2009.5354116.

Spigel, L. 2001. "Media Homes: Then and Now." *International Journal of Cultural Studies* 4(4): 385-411.

Spigel, L. 2005. "Designing the Smart House: Posthuman Domesticity and Conspicuous Production." *European Journal of Cultural Studies* 8(4): 403-26. doi: http://dx.doi.org/10.1177/1367549405057826.

Steinfeld, A., T. Fong, D. Kaber, M. Lewis, J. Scholtz, A. Schultz, and M. Goodrich. 2006. "Common Metrics for Human–Robot Interaction." Proceedings of the 1st ACM SIGCHI/SIGART conference on Human–robot interaction. 33-40. doi: http://dx.doi.org/10.1145/1121241.1121249.

Tay, B., Y. Jung, and T. Park. 2014. "When Stereotypes Meet Robots: The Double-Edge Sword of Robot Gender and Personality in Human–Robot Interaction."

Computers in Human Behavior 38 :75-84. doi: http://dx.doi.org/10.1016/j. chb.2014.05.014.

Tay, B., T. Park, Y. Jung, Y. Tan, and A. Wong. 2013. "When Stereotypes Meet Robots: The Effect of Gender Stereotypes on People's Acceptance of a Security Robot." In *Engineering Psychology and Cognitive Ergonomics. Understanding Human Cognition*, 261-70. Springer. doi: http://dx.doi.org/10.1007/978-3-642-39360-0_29.

Thrift, N. 2003. "Closer to the Machine? Intellegent Environments, New Forms of Possession and the Rise of the Supertoy." *Cultural Geographies* 10(4): 389-407. doi: http://dx.doi.org/10.1191/1474474003eu282oa.

Thrift, N. 2004. "Electric Animals: New Models of Everyday Life?" *Cultural Studies* 18(2-3): 461-82. doi: http://dx.doi.org/10.1080/0950238042000201617.

Torrey, C., A. Powers, M. Marge, S.R. Fussell, and S. Kiesler. 2006. "Effects of Adaptive Robot Dialogue on Information Exchange and Social Relations." Proceedings of the 1st ACM SIGCHI/SIGART conference on Human–robot interaction. 126-33. doi: http://dx.doi.org/10.1145/1121241.1121264.

Turkle, S., C. Breazeal, O. Daste, and B. Scassellati. 2006. "Encounters with Kismet and Cog: Children Respond to Relational Artifacts." In *Digital Media: Transformations in Human Communication*, edited by Paul Messaris and Lee Humphreys, 313--330. New York: Peter Lang Publishing.

Turkle, S., W. Taggart, C. Kidd, and O. Daste. 2006. "Relational Artifacts with Children and Elders: The Complexities of Cybercompanionship." *Connection Science* 18(4): 347-61. doi: http://dx.doi.org/10.1080/09540090600868912.

Wajcman, J. 2000. "Reflections on Gender and Technology Studies: In What State Is the Art?" *Social Studies of Science* 30(3): 447-64. doi: http://dx.doi.org/10.1177/030631200030003005.

Walters, M.L., D.S. Syrdal, K.L. Koay, K. Dautenhahn, and R. Te Boekhorst. 2008. "Human Approach Distances to a Mechanical-Looking Robot with Different Robot Voice Styles." Robot and Human Interactive Communication, 2008. RO-MAN 2008. The 17th IEEE International Symposium on. 707-12. doi: http://dx.doi.org/10.1109/ROMAN.2008.4600750.

Wang, Y. and J.E. Young. 2014. "Beyond Pink and Blue: Gendered Attitudes Towards Robots in Society." Proceedings of Gender and IT Appropriation. Science and Practice on Dialogue-Forum for Interdisciplinary Exchange. 49.

Weber, J. 2005. "Helpless Machines and True Loving Caregivers: A Feminist Critique of Recent Trends in Human–Robot Interaction." *Journal of Information, Communication and Ethics in Society* 3(4): 209-18. doi: http://dx.doi.org/10.1108/14779960580000274.

Weber, J. 2008. "Huamn–Robot Interaction." In *Handbook of Research on Computer-Mediated Communication*, edited by Sigrid Kelsey and Kirk St. Amant, 855-63. Hershey, PA: Idea Group Publisher.

Weber, J. 2014. "Opacity Versus Computational Reflection: Modelling Human–Robot Interaction in Personal Service Robotics." *Science, Technology & Innovation Studies* 10(1): 187-99.

Weber, J. and C. Bath. 2007. "'Social' Robot & 'Emotional' Software Agents: Gendering Processes and De-Gendering Strategies of 'Technologies in the Making'." In *Gender Designs It: Construction and Deconstruction of Information Society Technology*, edited by Isabel Zorn, Susanne Maass, Els Rommes, Carola Schirmer and Heidi Schelhowe, 54-63. Weisbaden: Verlag für Sozialwissenschaften.

Willson, M.A. 2012. "Being-Together: Thinking through Technologically Mediated Sociality and Community." *Communication and Critical/Cultural Studies* 9(3): 279-97. doi: http://dx.doi.org/10.1080/14791420.2012.705007.

Winner, L. 2010. *The Whale and the Reactor: A Search for Limits in an Age of High Technology*. Chicago, IL: University of Chicago Press.

Zhao, S. 2006. "Humanoid Social Robots as a Medium of Communication." *New Media & Society* 8(3): 401-19. doi: http://dx.doi.org/10.1177/1461444806061951.

Złotowski, J., D. Proudfoot, K. Yogeeswaran, and C. Bartneck. 2014. "Anthropomorphism: Opportunities and Challenges in Human–Robot Interaction." *International Journal of Social Robotics*:1-14. doi: http://dx.doi.org/10.1007/s12369-014-0267-6.

Chapter 12

Persuasive Robotic Technologies and the Freedom of Choice and Action

Michele Rapoport[1]

Automated, intelligent, persuasive technologies are permeating daily life and making their appearance felt in homes, cars, and workspaces. These forms of social robotics are designed to ensure that certain actions take place and that sought-after effects are attained; to do so they direct users towards making 'correct' decisions or by performing the 'right' actions themselves according to pre-programmed protocols. Engaging in habitual tasks, seemingly the operation of these devices is of little ethical concern; and yet, offering familiar albeit perhaps more efficient mechanisms of self-regulation and normalization, the deployment of these devices both challenges and reshapes users' abilities to choose and to act freely. Consequently, not only does personal freedom in using these devices need to be reassessed, but as these devices ascertain that 'the right thing gets done' they infringe upon personal autonomy and the constitution of subjectivity. Through their recontextualization of freedom and their impact on users as moral agents, these intelligent, persuasive technologies shape a new ethical landscape bound to action, habit, and device.

Introduction

> Bind me
> Hand and foot upright in the mast-step
> and tie the ends of the rope to the mast.
> If I command you and plead with you
> To release me, just tie me up tighter ... (Homer 2000, verses 167-71)

In his struggle to resist the enticing song of the sirens, Odysseus ordered his men to bind him to the ship's mast and seal their ears with wax so that none would succumb to the temptations of the Singing Sirens. Similarly, in their promise to "effectively encode experiences that change behaviors" (Fogg 2009), many

1   I am exceedingly grateful to the senior and junior fellows at the Edmond J. Safra Centers for Ethics at Tel Aviv University and at Harvard University for their insightful and instructive comments on previous drafts of this chapter.

intelligent technologies currently entering the market promise to resolve this very human of conflicts by allowing users to program their systems and devices to steer them towards the right decisions, or to make certain that proper outcomes are reached. Social robotics thus include many devices that are concerned with behavior and action; their raison d'être is to attain desirable results, to evoke certain performances, and to ensure that certain procedures are enacted that will bring about sought-after consequences. They are programmed to instate 'correct' and 'proper' behavior; behavior which at times may conflict with choices that otherwise might have been made by their users.

No longer solely the prerogative of science fiction, intelligent persuasive devices are already affecting the lives of contemporary individuals and communities, and as such require us to question the implications that their use will have on an emerging techno-human condition. This study will focus on if and how such devices that are voluntarily installed and deployed by their users and which take on persuasive capabilities affect possibilities for free choice and free action. Rather than posit that smart technologies are merely a more efficient and comprehensible illustration of technology's instrumentality or, alternatively, that they suggest a technological experience that is ontologically, and hence ethically altogether new—these technologies educe changes in possibilities for free choice and action that are both of degree and of kind, both quantitative and qualitative. Thus, while in many respects they offer a familiar, albeit perhaps more efficient mechanism of self-regulation, they also partake in a reshaping of personal freedom fashioned by the new ways humans and intelligent technologies interact. As interfaces with social robotics take on new form, it behooves one to critically delve into the possible challenges and restrictions that these new, emerging technologies will pose for personal freedom, and to address their ethical implications. It is this study's objective to attempt just that.

Intelligent and Persuasive Technologies

The 20th century saw the birth of robotics and various forms of intelligent technology, characterized by the ability to gather information on the environment, to process that information in order to draw intelligent inferences from it, and to act on those inferences by changing its performance. These devices are context aware, personalized, anticipatory, and adaptive; they create an interactive, computational sensorium that facilitates the symbiotic interaction between humans, devices, and their environments. As they establish networks of connectivity, they oversee the operation of appliances, systems and devices and coordinate their co-operation to create a synchronized technological setting. In addition to offering greater efficiency, ease, comfort, and productivity, they can be assigned numerous tasks, such as motivating and supporting healthier lifestyle choices or encouraging the enactment of various social norms and prescriptions. Many of these devices therefore assume persuasive capabilities as they direct users towards preferred and predetermined modes of action. Thus, for example, recent decades have seen devices that assist users in overcoming smoking habits, smart refrigerators that monitor food consumption and promote healthy eating habits, bathroom scales that monitor weight, and devices that restrict the consumption of energy and natural resources.

The persuasive capabilities of many smart devices have garnered much interest among researchers such as B.J. Fogg of the Persuasive Technologies Lab at Stanford University, who describes these emerging technologies as "fundamentally about learning to automate behavior change" (ibid.). The persuasive capacities of these devices can take on different forms: I am in a rush but cannot exceed the speed limit because my hybrid car prevents me from doing so, or I crave ice cream but my smart refrigerator, knowledgeable about my diabetes, has refrained from purchasing sugar-laden items. When installed by users, many of these devices partake in deliberate self-imposed processes of self-compulsion and self-regulation. The technological environment they create continuously enacts the physical, social, and mental choices of its users, whether by directing them to act in certain ways or by performing certain acts on their behalf.[2]

Mechanisms of Self-Regulation

> ... moral liberty, which alone makes man truly the master of himself. For to be driven by appetite alone is slavery, and obedience to the law one has prescribed for oneself is liberty. (Rousseau 1987, p. 27)

In thinking about various mechanisms employed to foster processes of self-control, self-management, self-regulation and self-limitation, one encounters the inherent binarism embedded in these actions and the state of conflict they manifest: I want that which I ought not to want; my desires go against my best interests. They posit the individual as one who may succumb to acrasia, namely to acting against his/her own better judgment, and to conflicts of a temporal dimension (long term versus short term), different mental facilities (rational versus emotive) or conflicting moral commitments (to myself versus to others). What they embody is not an ambiguous or relativistic framework, but rather one in which choices can have either positive or negative value. There is, ultimately, a *correct* decision to be made and an ensuing *correct* action to be performed. In choosing incorrectly one has, in many respects, failed one's self. Since these self-regulating mechanisms are inherently bound to action and performance, success entails the correct act being performed. The concern, therefore, is not whether one *wants* to exercise

---

2   This study focuses on devices tied to physical action or lack thereof, so smart online services such as '*Freedom*', the internet blocking productivity software program, or entertainment-related devices such as TeVo, which do not affect the ways in which humans move and behave in real, rather than virtual space will not be addressed.

but rather whether one has actually done so. In Frankfurt's terms, what matters are the 'effective' desires that in reality issue action and are acted upon; these are, ultimately, the manifestations of will (1971).

A situation of internal conflict manifests a juxtaposed corporeal or incorporeal 'other' who steers one towards the proper outcome. Reason—the capacity to question the desirability of being moved by impulses and desires—assumes ascendency and authoritativeness to facilitate a being divided from itself: "If I am to govern myself there must be two parts of me, one that is my governing self, my will, and one that must be governed and is capable of resisting my will" (Korsgaard 2008, p. 60). When a chocoholic sets the chocolate on a difficult-to-reach shelf or a smoker refrains from buying cigarettes, they are purporting a division within themselves wherein their desirous, potentially destructive impulses are overridden by corrective capacities. These join self-imposed restrictions that engage other individuals through the delegation of decision-making processes and/or performance: please remind or encourage me to do this or avoid that, please hinder me from acting in a certain manner. Similarly, material objects may be used to promote processes of self-regulation and self-management: alarm clocks, body scales, date book planners and blood pressure monitors—are used to further desirable outcomes.

Smart devices perform in a similar manner and will, according to manufacturers, do so in an efficient and uninterrupted manner. In the *Smart Kitchen for Nutrition-Aware Cooking*, for example, the 'Calorie Display' feature joins the 'Nutritional Facts Display' feature to gear users towards the consumption of nutritionally balanced meals. As initial observations have shown, the proposed smart device leads participants to exhibit different cooking behavior as they adjust ingredients to meet prescribed caloric counts (Jen-Hao et al. 2010). This technology promises to help users make more informed decisions while it provides them information on what they *are* doing in comparison to what they *ought to be* doing. Similarly, a 2002 patent describes "a monitoring device which can monitor automatically and in real time the electric consumption in a house, in premises or any other building while controlling the total consumption of the complete electric installation in a predetermined area" (from the abstract section of Rodilla 1998). Users of this device can control consumption both when present or absent, and limit total energy usage and costs.

While some smart technologies are tailored to counter individual preferences and tendencies of singular users, others profess a socially prescriptive dimension. Not mutually exclusive, however, some devices do both. Smart lock-down mechanisms that lock all entries into homes and offices and smart laundry devices that notify residents when the laundry must be cleaned assist users in overcoming their tendency to procrastinate, and in this respect serve as a means of overcoming specific and localized concerns harbored by users. Yet there are devices that instate norms and values more social by nature that impact the seemingly personal choices that individuals make. The conservation of energy resources, for example, is posited not only as economically beneficial to the individual or family, but bound to social liability and to encouraged sustainability practices. Devices that do so ensure that users will not only exercise fiscal prudence and cut personal costs, but that they also partake in processes of responsible citizenship. In this socially prescriptive dimension—one which directs towards actions with larger social implications—smart technologies join an array of what have been termed libertarian paternalism or 'nudge' mechanisms.

The term 'nudge' as positive reinforcement used to attain non-forced compliance has been adopted by Thaler and Sunstein to describe "any aspect of the choice architecture that alters people's behavior in a predictable way without forbidding any options or significantly changing their economic incentives" (2008, p. 6). With soft paternalism and libertarian paternalism as its underlying rationale, nudge helps people to make the 'right' and 'sensible' decisions without seemingly restricting their freedom. Criticized as leading to behavior desired by the state and serving as an economic policy instrument with impact on decision-making, as well as a form of mental manipulation that affects personal autonomy, nudge nonetheless remains what is perceived as a justifiable means of protecting people from themselves in cases where they are not acting in their best interest as, for example, in matters pertaining to health (see Schnellenbach 2012, Hausman and Welch 2010, Thaler and Sunstein 2003). Examples include an increase in the consumption of low-fat foods when these are most easily seen and reached in the layout of high school cafeterias, and wearable devices that monitor movement in order to promote increased physical activity. While health is clearly of primary concern to most individuals, nudging towards health also has social and economic ramifications affecting the number of employee sick-days, insurance expenditure, and allocation of resources for institutional health care. Criticized for exploiting irrationality and relying on human inertia and weakness of will (Bovens 2009), nudge techniques raise questions as to how choices actually reflect preferences, and in cases where preferences do change—how much is attributable to nudging and how much to the capacity to grow and learn.

When asking whether smart devices with nudge-like capabilities adopted for self-regulation affect one's ability to choose and act freely, one answer may be found in Frankfurt's categorizations of freedom of action and freedom of will, and in his understanding of how humans can have freedom of will even when their freedom of action is affected. In *Freedom of the Will and the Concept of a Person*, Frankfurt (1971) begins by distinguishing between effective (action-causing) and ineffective desires. He similarly differentiates between first order and second order desires: while the first entails a desire for anything other than a desire, the latter is the desire for a desire. So, for example, while one may not desire to take a morning run (first order), one may, nevertheless, desire to desire to exercise (second order). If we want this desire to be our will and the decision were made to nonetheless exercise and abide by second order desires, this would become an effective desire, defined as a *second order volition*.

For social robotics and intelligent devices, programming reflects second order desires, namely a desire to desire (I want to be fit, I want to take my medication on time); these devices render desires effective once they are put into action. I may want to want to be fit in general though at any given moment would prefer not to exercise. In this case, a persuasive device that encourages me to overcome my reluctance helps me fulfill my second order desire. Nevertheless, nudged by my smart device I may ultimately exercise not because my second order desire has been effective but because my device has prompted me to do so. Similarly to Frankfurt's example of the man who concentrates on his work not because of his second order desire to do so but due to some other, unknown motive, I may set out on my morning run not as a fulfillment of my second order desire to stay fit but rather because my device is vibrating annoyingly or because if I fail to exercise I will be interfering with a sequence of other ensuing events. If actions stem from desire, one has, according to Frankfurt, acted freely. One thus acts freely when the desire on which one acts is that which one desires to be effective. As this applies to persuasive devices, even if at the moment of action one has been manipulated or even coerced into performing in a certain manner, one has nonetheless acted freely if actions correspond to second order desires.

"Just as the question about the freedom of an agent's action has to do with whether it is the action he wants to perform, so the question about the freedom of the will has to do with whether it is the will that he wants to have" (Frankfurt 1971, p. 15). According to Frankfurt, in addition to freedom of action one also enjoys a freedom of will when effective desires are free and can be controlled. Freedom of will, therefore, entails that one has second order volitions and that first order desires correspond to them. A drug addict, therefore, has no free will as long as her second order volition fails to control her effective desires and ultimately her actions, and a dog has no free will because he only desires to eat but has no desire to desire to eat. In the current context, one has free will even when acting according to the persuasion of a technological device, since this suggestion fulfills the desire that one desires to fulfill. Interestingly, persuasive technologies affect the effectivity of first order desires, since if an effective desire is one that issues action, these devices may bring about the enactment of actions that may not correspond to first order desires. One may, for example, desire to eat cookies but end up eating vegetables not because second order desires to stay healthy and fit have taken effect, but because technology has barred one from acting upon one's initial, detrimental desire.

It may be that persuasive devices do not mark a paradigm shift but offer a more efficient means of enforcing second order desires; their 'set it and forget it' capabilities free one from continued deliberations on the micro level, at each singular point when decisions must be made. One could therefore argue that persuasive devices involve liberation from the tyranny of base desires, making way for desires for the good. Indeed, Frankfurt's classification of desires into effective and second order desires acknowledges a binary distinction between a controlling faculty on the one hand, and desires that ought to be restrained on the other. Through this distinction he constructs a framework where, albeit the presence of a dominant controller, the unified subject is exercising the capacity to will freely even in situations when he is unable to act freely.

However, counter to a perspective that sees no contradiction between these technologies and free will, it may also be claimed that it is at the very minute level of deliberation, at the moment when one is called upon to correlate first and second order desires, that freedom of will is exercised. If "[i]t is in securing the conformity of his will [*namely his effective first order desires*] to his second-order volitions, then, that a person exercises freedom of the will" (ibid.), than by not having to secure this conformity at the *ad hoc* level free will, in the more pedestrian activities of daily life, becomes inconsequential. Yet, as will be later argued, there are ethical implications to the setting aside of free will in the enactment of habitual activities, and that it ethically matters how much our actions are the result of a correspondence between what we want, and what we want to want.

Technologies of the Self

The confluence of the individuated and social dimensions of intelligent devices that perform as self-nudging mechanisms merits further investigation. As opposed to serving as a means of resolving 'internal' conflict, the societal dimension of many persuasive devices nudges one towards that which *should* or *ought* to be done within the larger social context. As such, technology becomes a means of ensconcing values and norms as it draws the public into the private sphere and imprints it upon the body.

> The alarm sounds at 7:00, which signals the bedroom light to go on as well as the coffee maker in the kitchen. The inhabitant, Bob, steps into the bathroom and turns on the light. The home records this manual interaction, displays the morning news on the bathroom video screen, and turns on the shower. While Bob is shaving, the home senses that Bob is four pounds over his ideal weight and adjusts his suggested daily menu that will later be displayed in the kitchen ... During breakfast, Bob requests the janitor robot to clean the house. When Bob leaves for work, the home secures all doors behind him and starts the lawn sprinklers. When Bob arrives home, his grocery order has arrived, the house is back at Bob's desired temperature, and the hot tub is waiting for him. (Cook and Yongblood 2004, p. 623)

The capacity of these devices to steer Bob towards his preferred choices is unmistakable: there is a routine to be adhered to, deadlines to be met, chores to be performed, personal, and environmental hygiene to be maintained and bodily measurements to be monitored and corrected. While this is ostensibly a perfectly tailored environment designed to meet all of Bob's singular and unique preferences, there is a normalizing dimension to it as well. One is expected to conserve natural resources, adhere to a healthy lifestyle, and take proper medication; divergence from these prescriptions is considered irresponsible at the least and subversive at the most. Bob is enveloped within an intelligent environment; his every action recorded, documented, and monitored, all decisions regarding the well-being of his body already made. His choices and preferences have been programmed into these devices and are to be continuously fed back to him. Ultimately, this is nudge on a grand scale—everywhere, in all things and all the time. And while this network of devices may provide Bob with an environment that is the epitome of comfort and convenience for him, in its prescriptive dimension it becomes a normalizing and disciplinary setting that profoundly impacts him. The proliferation of intelligent devices facilitates the amplified penetration of the public sphere into his personal and private choices and actions. While the settings that Bob programmed into his devices are underscored as personal and private choices (because who would not opt for a clean, healthy, and financially sensible lifestyle?), the continuous regulation of his body and its immediate surroundings sets it as an "object and target of power" that may be "subjected, used, transformed and improved" through coded activities and prescribed movements (Foucault 1995, p. 136).

According to Foucault, the body becomes a site where regimes of discourse and power inscribe themselves. In *Discipline and Punish* (1995) he describes the examination and rearrangement of the body and the birth of disciplinary power mechanisms in the 17th and 18th centuries. Employing numerous techniques including enclosure in singular, partitioned spaces (classrooms, barracks), the division of time into units dedicated to certain activities, the imposition of certain gestures (kneeling while praying, army drills), and specific relations between bodies and singular objects (guns, pens)—disciplinary mechanisms assume control over the activities, operations, and positions of docile bodies. It is in this manner, according to Foucault, that the regulation of all modes of sleeping, eating, praying, and speaking in monasteries facilitate the transformation of men into monks, and the repetitive routines, structured activities and regulations on marching, dressing, and bathing transform recruits into soldiers.

Social directives are already embedded in the very choices intelligent technologies are intended to facilitate: one cannot choose to install an energy-wasting or medicine-forgetting device, since such devices do not exist. At the initial planning, design, and funding stages societal values are already inscribed in these intelligent societal robotics, designed to affect decision-making under the guise of appealing to rational, ethical or even frugal aptitudes. As a means of instantiating norms and values, they assume disciplinary capacities not only *regardless* of the fact that it is the users themselves who have installed, programmed, and operated these devices, but also through the very decision to install them in the first place.

While devices intended to influence decision-making render processes of self-discipline and self-regulation more accessible, user-friendly and pervasive, their effect, on the whole, is still more in the realm of quantity than quality.[3] Consequently,

---

3   Though an argument may be made that a quantitative change may ultimately bring about a qualitative one. When the surveillance of employees increases in the work space

criticism directed towards these technologies as nudging or coercing mechanisms is similar to that directed at their 'non-intelligent' counterparts. As they take on nudge-like capabilities of which their users may or may not be aware of, concerns arise regarding the freedom to choose and act accordingly, and regarding behavior modification. Additional apprehensions concerning the employment of nudge and disciplinary tactics pertains to the formation of a morally lazy, fragmented self, in cases where one chooses to act differently when nudged as opposed to when one is not (Bovens 2009). Thomas Nagel's argument, moreover, addresses the modification of others' behavior and warns about sidestepping the biases that have lead one to make certain choices, thus failing to address the foundations of decision-making that govern certain decisions (2011).

Outcome-Oriented Devices

In addition to intelligent technology's capacity to affect decisions and choices as to how to act, many of these devices also act on the user's behalf, interceding at the performative level in order to insure final outcomes and results. Rather than suggest, advise or recommend, they ensure that the 'correct' act is performed. In the words of leading researchers in intelligent, automated driving: "[t]he driver is still in control. But if the driver is not doing the right thing, the technology takes over" (Markoff and Sengupta 2013). When smart doors allow admission to some individuals but prohibit access from others, when the temperature of the water in the shower can go no higher than a prescribed degree, or when emergency services are automatically contacted in case of an accident—what is imperative is the final outcome rather than the process by which it was reached. The question that arises is do such outcome-oriented devices suggest new paradigms for human-technology interaction and reinscribe possibilities for free choice and action as they, rather than nudge, coerce or manipulate users to perform desirable actions, instead compel them to act in certain ways and ascertain that certain outcomes are played out?

For outcome-oriented devices *ad hoc* decision-making is irrelevant as one is compelled to perform in a certain way—to reduce driving speed, for example. Similarly, when these devices operate in lieu of their users they do so according to input from the environment, which they continuously read, and not as a result of impromptu decisions made by users. The only real spontaneous choice is to go against the device itself. These technologies, therefore, mark an exchange: rather than making singular and routine decisions, users who install them are faced with

---

through the implementation of various sensors, programs, and monitors that document productivity and efficiency, changes in employer-employee relations are more than that of degree.

two options: surrender free choice and do as the device suggests, or disengage from the device altogether.[4]

Adopting Frankfurt's notion of passive action (1978), even when users are seemingly freed from the need to perform certain actions (with technology interceding on their behalf) they nonetheless remain agents of these actions:

> A driver whose automobile is coasting downhill in virtue of gravitational forces alone may be entirely satisfied with its speed and direction, and so he may never intervene to adjust its movement in any way. This would not show that the movement of the automobile did not happen under his guidance. What counts is that he was prepared to intervene as necessary and that he was in a position to do so more or less effectively. (ibid., p. 160)

The ability to effectively intervene—even to the point of overriding the device—establishes users of these devices as agents of action and hence as morally accountable for the outcomes of performances carried out in their stead.

Indeed, in *Where Are the Missing Masses? The Sociology of a Few Mundane Artifacts*, Bruno Latour (1992) describes his attempts to avoid putting on his seatbelt by overcoming the automated alarm feature:

> Early this morning, I was in a bad mood and decided to break a law and start my car without buckling my seat belt. My car usually does not want to start before I buckle the belt. It first flashes a red light 'FASTEN YOUR SEAT BELT' then an alarm sounds; it is so high pitched, so relentless, so repetitive, that I cannot stand it. After ten seconds I swear and put on the belt … A law of the excluded middle has been built, rendering logically inconceivable as well as morally unbearable a driver without a seat belt. Not quite. Because I feel so irritated to be forced to behave well that I instruct my garage mechanics to unlink the switch and the sensor. The excluded middle is back in! There is at least one car that is both on the move and without a seat belt on its driver—mine. (1992, pp. 125-6)

As this quote exemplifies, one is compelled to act in a certain way at the insistence of technology, and choosing not to do so becomes an act of defiance. As it takes on prescriptive capabilities, the warning feature in Latour's car assumes a form of moralizing agency as it advocates preferable behavior. One could only assume that the next step in the development of automated smart cars will be safety belts that strap one in and open automatically, adjusting to predetermined tightness and overall comfort settings and responsive to changing environmental conditions. In

---

4   There are, however, repercussions to such decisions. Overriding smart security and access settings may impact insurance claims, and ignoring devices that remind patients to take certain medication may have negative consequences in future dealings with health provider services.

these cars overcoming technology will become a much more dramatic gesture; it will become not an act of resistance but of insurgence.

Herein lies a threat to personal autonomy, for if autonomy is defined as "some ability both to alter one's preferences and to make them effective in one's actions" (Dworkin 1988, p. 108), than smart devices render *ad hoc* preference changes problematic (as one must override preset operational programs in the device) and, more importantly, they weaken the correlation between preference and ensuing action (since I may not need to act on my preferences but have the device do it for me instead). If "I am autonomous if I rule me, and no one else rules I" (Feinberg 1973, p. 161), than the proliferation of smart devices serves to undermine one's personal sovereignty as it introduces a range of social discourses that 'rule' one as well.

And yet, even when technology compels certain behavior or enacts it in our stead, even when it demands that we forcefully go against it rather than merely ignore its suggestions, we are still interacting with technology as subjects who make use of objects in order to concretize will. Many of the questions that persuasive social robotics evoke regarding one's ability to choose and act freely, while amplified, nonetheless remain within the same paradigmatic framework as those pertaining to nudge, coercion, and even substitution mechanisms. However, as the final chapters of this study would posit, it is in their capacity to integrate themselves into the most mundane, habitual, and often intimate actions of daily life that persuasive technologies promulgate a profound shift in the nature of freedom, affecting possibilities for resistance and shaping subjectivity. As they permeate the mundane, habitual, and often intimate settings of daily life, they encroach on and ultimately hinder possibilities for resistance; in their capacity to influence action and result they cast a new role for technology in the shaping of the free subject and impose certain subjectivities while compelling users to resist others.

The Freedom of the Subject

For Foucault, disciplinary power does not simply train bodies to comply with its regimes, but also produces forms of embodiment that lead to the constitution of subjectivity. The docile bodies that Foucault describes are required to perform in specific ways in the smallest of gestures, and in so doing become subjects engaged in ongoing process of self-examination, continuously measuring and comparing themselves against prevailing norms. These 'practices of the self' through which "the subject constitutes itself in an active fashion" are not "something invented by the individual himself. They are models that he finds in his culture and are proposed, suggested, imposed upon him by his culture, his society, and his social group" (Foucault 1997, p. 291). Taking this notion of subjectivity as grounded in daily practices and contextually shaped, *performativity*, as applied in philosophical, linguistic, and gender discourses, describes how what one does ultimately constitutes who and what one *is*.

In light of this understanding of performativity, interactions with persuasive technologies not only augment certain forms of subjectivity—citizen, homeowner, driver, employee, patient—but also constrict one's ability to challenge the enactment of disciplinary practices as they prompt certain performances rather than others. In Foucaultian terms, one could characterize intelligent devices as obstructing resistance—the manifestation of liberty—and hindering the emergence of new forms of subjectivity.[5] The struggle against "that which ties the individual to himself and submits him to others in this way (struggles against subjection, against forms of subjectivity, and submission)" (Foucault 1982, p. 781), is circumscribed when technology continuously governs performance and leaves little room for alteration and change, little room to 'refuse what we are' as subjects. As the smart car compels one to drive safely, park mindfully and consume gas responsibly, there is no choice but to be the dutiful driver one is expected to be. It is because these technologies are so deeply engaged in so many of the habitual and routine practices of daily life and because they are both temporally and spatially ubiquitous, that their disciplinary capacities recede from consciousness, rendering resistance all the more elusive. Adopting Foucault's examples, one can resist participating in a military drill or in the posturing of the body at prayer; yet resisting how one shops or bathes seems inexorably futile. For devices voluntarily installed by their users, resistance is all the more difficult (it was our choice, after all), and even if one were inclined to resist and perform differently, how is one to resist a practice that is enacted on one's behalf?

While it may be contended that smart technologies do not compel one to be a driver-subject, only to be a more cautious and mindful one, how are smart technologies to be understood *vis-à-vis* subjectivities that cannot be resisted? In the case of smart devices used, for example, in the homes of the elderly—devices that monitor and document data pertaining to the body, safeguard against accidents and remind users to take medication—age and its various manifestations are read as aberrant and therefore open to processes of normalization. The elderly body is deemed 'unnatural' and a viable target of corrective measures as the smart environment promulgates preferable practices. With these corrective measures available, choosing not to implement them may augment processes of social exclusion. If acting one's age or one's condition becomes a voluntary choice rather than a biological inevitability, choosing not to be 'improved' may be construed as an unwillingness to actively engage in an emerging technological condition. It becomes a choice with societal ramifications (necessitating perhaps more

---

5   The term 'resistance' is employed in its Foucaultian sense, namely "Where there is power, there is resistance, and yet, or rather consequently, this resistance is never in a position of exteriority in relation to power … Their [power relationships] very existence depends on a multiplicity of points of resistance … They are the other in the relations of power; they inscribe themselves as irreducible in relation to it" (Foucault 1978, pp. 95-6).

dependence on public funding for health, for example), and may be construed as a form of marginalization for which the subject is to be held accountable.[6]

Indeed, intelligent technologies are inexorably bound not only to the body, but to its weak and imperfect features. Addressing apprehensions towards mortality, they are concerned with its safety, health, and wellbeing; addressing tendencies towards sedentariness and indolence they offer efficiency, convenience, and physical comfort; addressing forgetfulness, distraction, and an absent-mindedness that may come with age they extend and materialize mental capacities. Thus, they not only augment and solidify certain subjectivities but, in a larger sense, perpetuate the ideal of what a proper, active, engaged human ought to be. In a world of intelligent technologies will one no longer be free to forget, ignore, procrastinate, age, or get sick? On a more banal level: when the Smart Closet suggests what to wear according to prevailing weather conditions will one not be robbed of the ability to miscalculate, err, and make mistakes? In assessing how personal freedom is reshaped through interactions with social robotics in general and persuasive ones in particular, there are, therefore, contradictory yet contemporaneous implications: on the one hand these technologies may grant greater personal freedom through the heightened autonomy they afford; yet on the other they may constrict one's freedom to accede to and comply with conditions of limitation, incapacitation, and exception.

Transhumanist visions in which technology aids humans in overcoming physical limitations and finitude through implants, modifications, and enhancements is echoed in the dream subsumed in intelligent technologies of transcending the flesh. As both body and environment are informated—approached through the lens of information and equated to it—they are predisposed to processes of interpretation and programming: "Change the code and you change the body" (Thacker 2003,

---

6   IIn her response to Sandel's *The Case Against Perfection* (2004), Kamm (2009) repudiates the former's rejection of enhancement as a form of mastery and an unwillingness to live with that which has been 'given', "a kind of hyperagency—a Promethean aspiration to remake nature, including human nature, to serve our purposes and satisfy our desires" (Sandel 2004). Sandel expresses a concern, which echoes that stated above, that technology may elicit the loss of humility *vis-à-vis* natural conditions and occurrences. Kamm's response that "it is not clear that there is anything morally preferable about normality at all" (2009, p. 109) is relevant to the current discussion, as is her rebuff of the correlation between choice and accountability. One can, she states, choose to forgo 'improvement' yet still be entitled to the same rights and benefits as those who would enhance both themselves and their surroundings. While Kamm is wary of romanticizing human failings—a wariness to which this author would subscribe – her approach may be far from the minds of service providers, such as insurance companies or security providers, who in their attempts to reduce expenditure will set cost reduction as their primary concern and make use of data collected on users to assign them fiscal accountability. If, for example, a user rejects the use of a smart medicine dispenser or refuses to install a smart security system, for these bodies the question is not whether this expresses acceptance or rejection of human vulnerability but rather, if by so doing, they will have brought about greater expenditure in the future.

p. 87). Yet while the possibility of standardizing the dynamic reality of human performativity implies systematization, even mechanization of human life, the personalization of intelligent technologies tailored to meet the particular needs of users suggests singularity and individuation. They extend their particular users' desires onto the environment and are granted a voice—figuratively and, occasionally, literally. Indeed, technologies become open to processes of anthropomorphization once perceived as speaking both to humans (telling us what to do) and to other devices through networked connectivity. Thus, concurrent with processes that mechanize the human are those that humanize the machine. Such an individuation of technology, it may be argued, entails further constrictions on one's ability to resist its dictates, since as devices are programmed to abide by individuated demands and requirements, it becomes all the more challenging to go against them, with such a move translating into a move against one's own, singularized extended will.

New Ethical Ground

> ... the nature of human action has changed, and since ethics is concerned with action, it should follow that the changed nature of human action calls for a change in ethics as well ... (Jonas 1984, p. 2)

This study posited the question of whether intelligent, persuasive devices are instrumental—and in this respect not ontologically new, or whether a fundamental shift in our abilities to choose and act freely will come about as these permeate the spheres of daily life. It has been shown that the nature of changes that these technologies propagate varies. They may perform similarly to various nudge mechanisms in their self-controlling capacities, but are also bound to social prescriptions, inscribing new forms of self-governance and manifesting a disciplining of the socially constructed subject. With the distinction being made between devices that urge one to act correctly and those that perform instead of their users, possibilities for free choice and action undergo alternative forms of reshaping and reinscription according to the different ways by which these devices operate.

Whether one focuses on the loss of negative or of positive freedom as defined in Berlin's dialectic of negative and positive liberty—in other words, whether one focuses on the absence of interference, obstacles or constraints that may hinder one's ability to act on the one hand, or on the evocation of personal autonomy through self-determination and control on the other—new limits on personal freedom effect a new ethical landscape that behooves examination (Berlin 1969). We find, on the one hand, persuasive devices that free one from the fallibility and vulnerability of the mortal, corporeal body (negative freedom) and grant heightened autonomy as they promise increased self-reliance through an environment that echoes and supports personal choices and preferences (positive freedom). On the other hand, these devices augment the presence of social dictates influencing the seemingly private choices humans make pertaining to their bodies and their environments and as such amplify the presence of societal norms and values and serve to constrict negative freedom. Similarly, they hinder one's ability to assert one's autonomy to choose freely as choices are predetermined and preprogramed into devices, ultimately inhibiting room for change, development or growth.

The final section of this chapter will attempt to examine if new ethical challenges have been shaped by the unique techno-human conditions that persuasive devices facilitate, and if so, what these may be. The first notable change that ought to be pointed out is a spatial one and pertains to a new contextualization of the discourse on freedom. Here the novelty lies in situating questions pertaining to free choice and action within the private rather than the public realm, and to the ways by which privacy and freedom are posited as oppositional and set against one another.

Privacy has traditionally been seen as a means of empowerment, allowing people to control the publication and distribution of information about themselves. It is utilitarian in that it enables one to repel unwanted intrusions while promoting the right to be physically alone. It is a means of reinforcing identity, independence and the boundaries of the subject. "Privacy is essential", states Hayles, "because it provides a space for resistance to the hegemonic power of states and corporations" (Hayes 2009, p. 313). Yet as persuasive, intelligent devices gather and store data on the intimate habits of consumption, hygiene, health, labor, personal safety and entertainment of their users, they produce a wealth of figures and facts; when such data is processed and interpreted to become useful, sensible, and meaningful, it becomes information (Introna 1997).

These new compilations of information made possible by technology move beyond digitalized data that quantifies reality and become that which affects the ways by which reality is perceived and constructed. Information, in other words, assumes an agential capacity.[7] As smart environments lead to the establishment of informated environments, not only are the boundaries of privacy undermined by the export of data beyond the confines of the private realm, but the creation of a new body of information goes towards shaping and reconstructing practices within individuated spaces. If, for example, information regarding the intake of medication or that pertaining to dietary habits is amassed, this body of information—this new source and form of knowledge—may compel users to alter their habits once medical insurance companies are also privy to it. Rather than private spaces being controlled by individuals and serving as sites of personal autonomy and freedom, informativity shapes the ways by which individuals perform their routine actions within their intimate spaces. Instead of serving as sites of reclusion and resistance, private domains, by generating a body of information, become sites of exposure and compliance.

---

7 For example salary information in the workplace that is kept confidential. The concern is that if publicly known this information will affect and necessitate change in workplace segregation and status attainment.

The recontextualization that takes place as new technologies permeate the spaces of daily life applies not only to freedom. These technologies also promote a re-siting of the ethical, so that the routine, mundane, prosaic actions that constitute daily life become augmented sites of ethical deliberation. Turning on the light, locking the door, deciding what to eat, how to drive, or remembering to take one's medication—these are not the dilemmas that ethics, in its traditional sense, has chosen to focus on. Rather, they are the mostly transparent actions, the negligible activities upon which the more notable ethical questions substantiate themselves. What is deemed significant is not whether we have locked the door to our home, office or car, but how our right to personal safety may infringe upon the rights of others, or to what extent we or the state are responsible for personal security. What has been shown, however, is that concerning ourselves with the alterations that intelligent technologies may exact upon possibilities to choose and act freely constitutes a re-siting of the ethical into arenas perceived as seemingly inconsequential. Our ostensibly insignificant actions, it appears, now draw the ethical eye; automatic actions, to which little thought was previously given, become sites of deliberation as we program our devices according to particular thought-out parameters or install certain devices that promise assured results (energy-saving, exercise-promoting, etc.). Attention is drawn to a range of daily activities once technological apparatuses are designed to invoke or replace them, shaping new fields of ethical deliberation: what are the ramifications of heating my home above the suggested temperature? Who will ultimately be accountable for the medical expenditures of over-eaters and under-exercisers? Am I to be denied medical coverage in case of an accident if I ignored the promptings of my smart car to fasten my seat belt?

As the habitual actions of daily life take on ethical significance, what are the implications of having technological devices in our private sphere decide and act for us? Similarly to nudge and coercive mechanisms, persuasive technologies have been shown to be situated at the interstices between action and desired outcome, between intention and result. Indeed, it is often challenging to correlate what one wants with what one actually does, and these devices manifest the attempt to automate and narrow the fissure that is inescapably there. Rather than perpetual struggles of having action correspond to intention, persuasive devices ascertain that the final deed is performed, though one may be lacking the motivation and moral impetus for its enactment. Users are spared from their own *ad hoc* ineptness at meeting self-prescribed goals, yet technology deprives them of the prospect not of 'doing the right thing' but rather of '*wanting* to do the right thing' at the point in time where intention leads to action.

In this respect, having technology steer users towards the right action or having it perform the right action in their stead shifts the individual's motives to the moral sidelines. While absence of a motivational structure may not necessarily impede on the final moral outcome, it can put into question the virtuousness of the enactor. Indeed, Aristotle draws this distinction between external performance and internal state, allowing for an action to be right but not necessarily virtuous: "The just and temperate person is not the one who does these [right] actions, but the one who also does them in the way in which just or temperate people do them" (Aristotle 1999, 1105b7-9). For Aristotle, virtue entails habituation, namely learning through practice to perform the proper actions (ibid., 1103b21-5), as well as enjoyment in performing these actions. Quoting Eunenus, Aristotle states: "Habit ... is longtime training ... and in the end training is nature for human beings" (ibid., 1152a33-5), so that if "a state [of character] results from [the repetition of] similar activities" (ibid., 1103b21-2), being nudged to do what is right or having technology perform the right action itself is not only *not* indicative of the presence of a virtuous enactor, but may also prohibit one from becoming virtuous with the virtuous act being neither voluntarily practiced nor enjoyed. Are we really inclined towards sustainability when we conserve energy or towards protecting our fellow drivers when abiding by our smart car's instructions? If we believe that meaning and intention matter, persuasive devices allows for a disparity with morally disconcerting implications.

When smart cars prohibit drivers from exceeding the speed limit or from driving while intoxicated, where is accountability to be assigned, particularly when unexpected, adverse events occur? Indeed, with limited personal autonomy are users still accountable for the actions made on their behalf by their technological devices?; Does being nudged into a certain course of action render users of persuasive technologies responsible to a lesser degree than they would have been without the use of these devices? One possible answer is offered by Floridi and Sanders (2004) in their conceptualization of an a-responsible morality that allots intelligent, outcome-oriented devices the capacity to induce the 'right' results, yet without responsibility for this induction. Their understanding of accountability in the use of smart devices distinguishes between moral agency and action agency, and their solution posits users as accountable even without having to perform ethical acts. The model they propose allows accountability to remain in the realm of the human while acknowledging that it is the technological device that facilitates the proper or correct outcome, in a sense allowing for a "moral agent [who is] not necessarily exhibiting free will, mental states or responsibility." A-responsible morality entails responsibility for actions even if one has not chosen to perform them, thus severing the link intuitively made between the ability to desire, choose, and act freely, and accountability for the enactment of these desires, choices, and actions.

Arguments have been made that Floridi and Sanders lower the threshold for moral agency (e.g. Coeckelbergh 2009), yet whether one does or does not subscribe to their distinction between responsible (human) and a-responsible (technological apparatus) morality, one nonetheless finds one's self engaged in an ethics that is beyond anthropocentrism and whose purview extends beyond human–human interactions. Social robotic persuasive devices not only serve as extensions of the will, but are a means of directing towards or enacting pre-defined modes of action. They become actively engaged in ascertaining that certain results are reached, and as such affect one's capacity to choose freely without being externally directed.

Moreover, they define the scope of deliberation, providing certain choices while eliminating others, and make it possible for users to avoid determining the right or wrong course of action at the very moment when action is required. Persuasive technologies are innately bound to moral discourses, albeit those which pertain to minute rather than momentous deliberations, and as such cannot be perceived a morally neutral or peripheral to ethical considerations.

References

Aristotle. 1999. *Nicomachean Ethics*. Translated by Terence Irwin. 2 ed. Indianapolis, Cambridge: Hackett Publishing Co.

Berlin, I. 1969. "Two Concepts of Liberty." In *Four Essays on Liberty*, 118-72. Oxford: Oxford University Press.

Bovens, L. 2009. "The Ethics of Nudge." In *Preference Change: Approaches from Philosophy, Economics and Psychology*, edited by Till Grüne-Yanoff and Sven Ove Hansson, 207-19. Dordrecht: Springer Netherlands. doi: http://dx.doi.org/10.1007/978-90-481-2593-7.

Coeckelbergh, M. 2009. "Virtual Moral Agency, Virtual Moral Responsibility: On the Moral Significance of the Appearance, Perception, and Performance of Artificial Agents." *AI & Society* 24(2): 181-9. doi: http://dx.doi.org/10.1007/s00146-009-0208-3.

Cook, D.J. and M. Yongblood. 2004. "Smart Homes." In *Berkshire Encyclopedia of Human-Computer Interaction: When Science Fiction Becomes Science Fact*, edited by William Sims Bainbridge, 623-7. Breat Barrington: Berkshire Publishing Group.

Dworkin, G. 1988. *The Theory and Practice of Autonomy, Cambridge Studies in Philosophy*. Cambridge: Cambridge University Press.

Feinberg, J. 1973. "The Idea of a Free Man." In *Educational Judgments: Papers in the Philosophy of Education*, edited by James F. Doyle. London and Boston: Routledge & Kegan Paul.

Floridi, L. and J.W. Sanders. 2004. "On the Morality of Artificial Agents." *Minds and Machines* 14(3): 349-79. doi: http://dx.doi.org/10.1023/B:MIND.0000035461.63578.9d.

Fogg, B.J. 2009. "A Behavior Model for Persuasive Design." Proceedings of the 4th International Conference on Persuasive Technology, Claremont, California, USA. doi: http://dx.doi.org/10.1145/1541948.1541999.

Foucault, M. 1978. *History of Sexuality*. Translated by Robert Hurley. Vol. 1. New York. Original edition, 1976.

Foucault, M. 1982. "The Subject and Power." *Critical Inquiry* 8(4): 777-95. doi: http://dx.doi.org/10.1086/448181.

Foucault, M. 1995. *Discipline and Punish: The Birth of the Prison*. Translated by Alan Sheridan. New York: Vintage Books.

Foucault, M. 1997. "Ethics: Subjectivity and Truth." In *The Essential Works of Michel Foucault 1954-1964, Volume 1*, translated by Robert Hurley, edited by Paul Rabinow. London: Penguin Press.

Frankfurt, H.G. 1971. "Freedom of the Will and the Concept of a Person." *The Journal of Philosophy* 68(1): 5-20. doi: http://dx.doi.org/10.2307/2024717.

Frankfurt, H.G. 1978. "The Problem of Action." *American Philosophical Quarterly* 15(2): 157-62.

Hausman, D.M. and B. Welch. 2010. "Debate: To Nudge or Not to Nudge." *Journal of Political Philosophy* 18(1): 123-36. doi: http://dx.doi.org/10.1111/j.1467-9760.2009.00351.x.

Hayes, K.N. 2009. "Waking up to the Surveillance Society." *Surveillance & Society* 6(3): 313-16.

Homer. 2000. *Odyssey*. Translated by Stanley Lombardo. Indianapolis, IN: Hackett Publishing Company.

Introna, L.D. 1997. "Privacy and the Computer: Why We Need Privacy in the Information Society." *Metaphilosophy* 28(3): 259-75. doi: http://dx.doi.org/10.1111/1467-9973.00055.

Jen-Hao, C., P.P-Y. Chi, C. Hao-Hua, C.C-H. Chen, and P. Huang. 2010. "A Smart Kitchen for Nutrition-Aware Cooking." *IEEE Pervasive Computing* 9(4): 58-65. doi: http://dx.doi.org/10.1109/MPRV.2010.75.

Jonas, H. 1984. *The Imperative of Responsibility: In Search of an Ethics for the Technological Age*. Chicago, IL: University of Chicago Press.

Kamm, F. 2009. "What Is and Is Not Wrong with Enhancement?" In *Human Enhancement*, edited by Julian Savulescu and Nick Bostrom, 91-130. Oxford: Oxford University Press.

Korsgaard, C.M. 2008. "The Normativity of Instrumental Reason." In *The Constitution of Agency: Essays on Practical Reason and Moral Psychology*, 27-68. Oxford: Oxford University Press. doi: http://dx.doi.org/DOI:10.1093/acprof:oso/9780199552733.003.0002.

Latour, B. 1992. "Where Are the Missing Masses? The Sociology of a Few Mundane Artifacts." In *Shaping Technology/Building Society: Studies in Sociotechnical Change*, edited by Wiebe E. Bijker and John Law, 225-58. Cambridge, MA: MIT Press.

Markoff, J. and S. Sengupta. 2013. "Drivers with Hands Full Get a Backup: The Car." *The New York Times*, January 12, Science. Accessed January 29, 2015. http://www.nytimes.com/2013/01/12/science/drivers-with-hands-full-get-a-backup-the-car.html?pagewanted=all.

Nagel, T. 2011. "David Brooks' Theory of Human Nature." *New York Times*, March 11, Sunday Book Review. Accessed January 29, 2015. http://www.nytimes.com/2011/03/13/books/review/book-review-the-social-animal-by-david-brooks.html?pagewanted=all&_r=1.

Rodilla, S.V. 1998. "Programmable Monitoring Device for Electric Consumption." Google Patents Accessed May 2015. http://www.google.com.ar/patents/WO1998050797A1?cl=en.

Rousseau, J-J. 1987. *On the Social Contract*. Translated by Donald A. Cress. Indianapolis, IN: Hackett Publishing Company.

Sandel, M.J. 2004. "The Case against Perfection: What's Wrong with Designer Children, Bionic Athletes, and Genetic Engineering." *The Atlantic*, April.

Schnellenbach, J. 2012. "Nudges and Norms: On the Political Economy of Soft Paternalism." *European Journal of Political Economy* 28(2): 266-77. doi: http://dx.doi.org/10.1016/j.ejpoleco.2011.12.001.

Thacker, E. 2003. "Data Made Flesh: Biotechnology and the Discourse of the Posthuman." *Cultural Critique* (53): 72-97.

Thaler, R.H. and C.R. Sunstein. 2003. "Libertarian Paternalism." *American Economic Review* 93(2): 175-9. doi: http://dx.doi.org/10.1257/000282803321947001.

Thaler, R.H. and C.R. Sunstein. 2008. *Nudge: Improving Decisions About Health, Wealth, and Happiness*. New Haven, CT: Yale University Press.

Index

Note: *Italics* indicate figures.

Abe, M. 105-6
action perception system (APS) 34n, 34-5
AI (artificial intelligence) hard/soft 85
AIBO (robot) 114, 201, 208n
Allen, C. 63
AMAs (artificial moral agents) 85
Ambo, P. 103, 114, *see also* PARO
    (robotic seal)
appearance, physical
    and empathy 30, 33, 35
    and ethics 102, 109
    and social order 158, 163-4, 164n
    and the uncanny valley phenomenon
        22-3, 126, 128
    see also computer graphics; gender
        codes in social robotics
APS (action perception system) 34n, 34-5
Arendt, H. 62n
Aristotle 63, 183n, 185, 234-5
artificial intelligence (AI), hard/soft 85
artificial moral agents (AMAs) 85
Austin, J.L. xx, 139, 146-7, 149, 150, 152
autonomy, human, *see* persuasive robotic
        technologies, and human freedom
        of choice and action

BA (bodymind awareness), *see* Buddhist-
        inspired thought on human–robot
        interaction (HRI)
Bach, M. 28n
Barnes, J.A. 143-4
Bartneck, C. 22n-23n, 30, 131, 203
Bath, C. 210
Becker, B. 58, 69
Bein, S. 112n
Berlin, I. 232
Billard, A. 201
Boden, M. 60, 61

body language, *see* understanding in
        human–robot interaction (HRI)
bodymind awareness (BA), *see* Buddhist-
        inspired thought on human–robot
        interaction (HRI)
Borenstein, J. 208
Brave, S. 210
Breazeal, C. 8-9, 14, 20, 36, 145, 201, 202
Bringsjord, S. 73n
Buddhist-inspired thought on human–robot
        interaction (HRI) 104-5, 108-13,
        111
    and bodymind awareness (BA) 106-8,
        107, 115-16
    and suffering 105-6
Burleigh, T.J. 32, 33
Butler, J. 139, 150-52

Callicott, J.B. 49
CAP (computing and philosophy) 59, 59n
Carpenter, J. 205, 208
Challa, P. 7
Clark, A. 60-61
Coeckelbergh, M. xvii, 43n-44n, 51, 52-3,
        60n, 66, 187, 204
cognitive capacities, *see* understanding in
        human–robot interaction (HRI)
companion robots 87-9, 164, 164n, *see*
        also PARO (robotic seal)
computer graphics 127-8
    capturing reality 129-30
    simulating reality 128-9
computing and philosophy (CAP) 59, 59n
consciousness, *see* understanding in
        human–robot interaction (HRI)

Danish Council of Ethics, The 101-2, 103
Dautenhahn, K. 7-8, 14, 200, 201
Debevec, P. 129-30

DeepMind 83-4
Défago, X. 187
dehumanization, xviii, xix, *111*, 111-12,
    118, 123, 133-4
Dennett, D.C. 35
Derrida, J. 139, 150
Descartes, R. 68
desire, *see* love, sexuality and the arts of
    being human, robots and
Digital Emily (digital clone) 129-30
Digital Ira (digital character) 130
DiSalvo, C. 202, 203, 208
discrimination, epistemology of, and
        human–robot interaction (HRI)
        99-101, 118
    and/as human self-realization 116-18
    Buddhist-inspired thought on 104-5,
        108-13, *111*
        bodymind awareness (BA) 106-8,
            107, 115-16
        suffering 105-6
    current public ethical debate 101-4,
        232-6
    technological dangers and potential
        113-15
    see also understanding in human–robot
        interaction; social world, and
        robots, humans
domestic robots, *see* persuasive robotic
        technologies, and human freedom
        of choice and action; theater,
        human–robot, and *I, Worker*
        (Hirata and Ishiguro)
Dostoyevsky, F. 43
Dreyfus, H. 60, 61
Duffy, B.R. 210

Eckersall, P. 149
eeriness, *see* uncanny valley phenomenon
Emily Project 129-30
emotions, *see* morality, robots and the
        limits of
empathy with sociable robots 19-21, 35-6
    and the uncanny valley 29-35, 30n,
        34n
    vs. eeriness 21-5
    on the imaginative perception of
        emotion 25-9

see also uncanny valley phenomenon
Enframing 113-14
eros, *see* love, sexuality and the arts of
        being human, robots and
ethics, current public debate 101-4, 232-6;
    see also human–robot interaction
        (HRI), and the epistemology of
        discrimination; love, sexuality and
        the arts of being human, robots
        and; moral and legal responsibility
        for robots; morality, robots and the
        limits of
Ethics Application Repository, The
        (TEAR) 91
ethics boards, for research in robotics and
        AI, 83-5, 94-5
    AI and robotic ethics 85-6
        long term ethical concerns 88-9
        short term ethical concerns 86-8
    choosing the right ethics 92-3
        codes of ethics, using and creating
            93-4
    purpose of ethics boards 89-92
Eyssel, F. 206, 207, 208

Fallon, A.E. 133
feelings, *see* morality, robots and
        the limits of
female robots, *see* gender codes in social
        robotics; love, sexuality and the
        arts of being human, robots and
Fincher, K. 133
Floridi, L. 165, 235
Fogg, B.J. 221
Fong, T. 200
Foucault, M. 226, 229-30, 230n
Fox, J. 190
Frankfurt, H.G. 222, 223-5, 228
freedom, human, *see* persuasive robotic
        technologies, and human freedom
        of choice and action
Freud, S. 123-4
Friedman, B. 114
friendship, *see* love, sexuality and the arts
        of being human, robots and
Frith, C. 112

Gallagher, S. 10

Gemperle, F. 202, 203
gender codes in social robotics 199-200
    affective and utilitarian social robots
        200-201
    anthropomorphism 201-4
    beyond gendered machines 210-11
    ethics and implications of gendering
        207-9
    human–robot interaction literature on
        the gendered social robot 204-7
Gerdes, A. 57n, 57-8, 62-4
Goetz, J. 202-3
Goodrich, M.A. 6
Google DeepMind 83-4
graphics, *see* computer graphics
Gray, D.M. 33-4
Gray, K. 33-4
Guiliani, M. 7
Gunkel, D. xvii, 51

Hagman, J. 114
Haidt, J. 133
Haley, A. 194
Haslam, N. 134
Hassabis, D. 83-4
Hataraku Watashi (theater production),
    see theater, human–robot, and *I,
    Worker* (Hirata and Ishiguro)
Haugeland, J. 60
Hayles, K.N. 233
Heerink, M. 201n
Hegel, F. 206, 207, 208
Heidegger, M. xix, xxi, 61, 68, 99, 100,
    101, 104, 113-18
Heisenberg, W. 108
Hellström, T. 185
hermeneutics 4, 4n, 10, 47
Hirata, O. xx, 6, 148, 149; *see also* theater,
    human–robot, and *I, Worker*
    (Hirata and Ishiguro)
Homer 219

I, Worker (Hirata and Ishiguro), *see* theater,
    human–robot, and *I, Worker*
    (Hirata and Ishiguro)
Iacoboni, M. 34n
Ihaya, K. 32-3
Ikemoto, S. 6

institutional review boards (IRBs) 86,
    90-91, 94-5
intelligent technology, *see* persuasive
    robotic technologies, and human
    freedom of choice and action
interaction, *see* human–robot interaction
    (HRI), and the epistemology of
    discrimination
IRBs (institutional review boards) 86,
    90-91, 94-5
Isbister, K. 209
Ishiguro, H. xx, 6, 31, 33, 108, 111,
    126-7, 131-2, *132*, 148, 207; *see
    also* theater, human–robot, and *I,
    Worker* (Hirata and Ishiguro)
Ishihara, K. 99n

Jakovljevic, B. 147, 149
James, B. 19n, 30n
Jentsch, E. 30-31, 33, 123
Johnson, D.G. 165

Kahn, P.H. 114, 208
Kamm, F. 231n
Kang, M. 23
Kant, I. 45n-46n, 63, 67, 71, 104, 165, 194
KASPAR (robot) 6
Kasulis, T.P. 106, 112
Kawabe, T. 32-3
Kiesler, S. 202-3
KISMET (robot) 103
Kitano, N. 103
Kosuge, K. 5-6
Kuflik, A. 185

Lacroix, G. 19n, 32, 32n
language, *see* understanding in
    human–robot interaction (HRI)
LARs (Lethal Autonomous Robots) 62-3,
    64
Latour, B. 228
law, *see* moral and legal responsibility for
    robots
Lethal Autonomous Robots (LARs) 62-3,
    64
Levinas, E. 65, 65n
Levy, D. 65, 72
Lindemann, G. 158, 161-2

LIREC project 145-6, 147-8
localizable robots, *see* moral and legal
    responsibility for robots
Looser, T.E. 33
love, sexuality and the arts of being human,
    robots and, 57-9, 74-6
  backgrounds 59-61
  good sex *vs.* complete sex, 67-72, 73n
  limitations of computers and robots
    61-5
  love and friendship, between humans
    and machines 72-4
  love and sex, beyond humans 65-7
  'robotic companions,' long term ethical
    concerns 88-9, 164, 164n
Lu, D.V. 146
Luxton, D. 87-8

MacDorman, K. 126-7, 131-2, *132*, 207
McDermott, D. 63
Maibom, H. 24-5, 25n, 27
male robots, *see* gender codes in social
    robotics
Matheson, D. 19n, 31n
Matloff, N. 84
May, R. 67
Merleau-Ponty, M. 61, 67, 68
Michaud, M.A.G. 194
Minato, T. 6
mirror neuron system (MNS) 34n
Misselhorn, C. xvii 19, 20-21, 22, 23,
    25n, 25-7, 26n, 28-30, 32, 35-6
MNS (mirror neuron system) 34n
moral and legal responsibility for robots
    177-9, 194-5
  current debate 184, 208n-209n
    as ersatz human being 184-6
    as localizable individuals or
      nonlocalizable communities
      186-7
    as a tool for human use, 184
  developing an ontological framework
    autonomy 180
    localizability 182
    purpose 179-80
    sources of nonlocalizability 182-3
    volitionality 181

frameworks for autonomous
    nonlocalizable robots, legal and
    ethical 192
  with animal-like autonomy and
    volitionality 192-3
  with human-like autonomy and
    metavolitionality 193
  with low autonomy and
    volitionality 192
  with superautonomy and
    supervolitionality 194
future, for nonlocalizable robots
  advances to the development of
    nonlocalizable robots 187-8
  as alien intelligence 190-91
  as animal other 189
  as human-like other 189-90
  as legal and moral leaders 191
  movement of 188-9
  see also ethics boards, for research in
    robotics and AI
morality, robots and the limits of 39-45,
    54, 61-5
  and environmental ethics 49-54
  non-derivative objects of moral
    consideration, robots as 45-8
  see also ethics boards, for research in
    robotics and AI
Mori, M. xix, 22-3, 22n-23n, 30, 124,
    124n, 131, 202, *see also* uncanny
    valley phenomenon

Nagel, T. 67, 227
Nass, C. 204, 207, 210
Natanson, M. 69
Niculescu, A. 206
Nishida, K. 111
nonlocalizable robots, *see* moral and legal
    responsibility for robots
Nørskov, M. 99n
Nourbakhsh, I. 200
nudge mechanisms 223-4, 225-7, 234-5

O'Brien, E. 129-30
ontology, *see* moral and legal responsibility
    for robots
Ovid 65

PARO (robotic seal) 103, 108-9, 114-15, 116, 117, 201, 201n, 208n

Pearson, Y. 208

persuasive robotic technologies, and human freedom of choice and action 219-20

20th century developments 220-21

freedom of the subject 229-32

mechanisms of self-regulation 221-5

new ethical ground 232-6

outcome-oriented devices 227-9, 228n

technologies of the self 225-7

phronesis 57-8, 61-5, 62n, 64n, 74-6

Plantec, P. 130

Plato, *Symposium* 65, 67

Pleo (robotic toy dinosaur) 20, 22

Polar Express, The (film) 125

Poulton, M.C. 149

Powers, A. 204-5

privacy 233

Pygmalion 65

realism, *see* uncanny valley phenomenon

Redstone, J. 21n

Reeves, B. 204, 207

responsibility, *see* moral and legal responsibility for robots

Riek, Laurel D. 144

Robertson, J. 141-2, 209

roboethics, *see* ethics, current public debate; ethics boards, for research in robotics and AI

Robot & Frank (film) 21-2, 30n, 36

Robotinho (robot) 6

Robovie (robot) 6

Rosenthal-von der Pütten, A.M. 24

Rousseau, J-J. 221

Rozin, P. 133

Ruddick, S. xvii 57, 58, 67-72, 70n, 75

Sandel, M.J. 231n

Sanders, J.W. 165, 235

Sartre, J-P. 67, 71

Saygin, A.P. 34-5

Schechner, R. 147

Schermerhorn, P. 205n

Schoenherr, J.R. 32

Scholtz, J. 6

Schütz, A. xvi, 3, 4n, 4-5, 10-14, 12n

Searl, J. 44

Seibt, J. 60n, 99n, 102n

Sekmen, A. 7

sex, *see* gender codes in social robotics; love, sexuality and the arts of being human, robots and

Shaner, D. xix, 106-8, *107*

Sharkey, A. 109

Shibata, T. 103, 108-9, *see also* PARO (robotic seal)

Siegel, M. 205

Smart, W. D. 146

smart devices, *see* persuasive robotic technologies, and human freedom of choice and action

social world, and robots, humans 157-9, 170-72, 208n-209n

bodily movements and actions of human–robot interaction (HRI) 10-14

interaction frames 166

objectified interaction/Ego-Alter-Tertius (Triad) 169-70

simple/Ego-Alter structure (Dyad) 166-9

levels of human–robot interaction (HRI) 5-9

problems of social order 159-66

Socrates 67

Sorensen, K.D. 46n

Sparrow, L. 100, 103, 109, 117

Sparrow, R. 100, 103, 109, 117

Spedalieri, F. 150

Stahl, B.C. 184, 186

Sullins, J. 57, 57n, 58, 64, 65-6, 67, 72-4

Sunstein, C.R. 223

Tay, B. 206

TEAR (The Ethics Application Repository) 91

technologies, robot, *see* human–robot interaction (HRI), and the epistemology of discrimination; persuasive robotic technologies, and human freedom of choice and action

Thaler, R.H. 223

theater, human–robot, and *I, Worker*
    (Hirata and Ishiguro) 139-40
  aesthetics of robot theater 147-9
  differentiation of citationality 150-51
  and identity crises 142-3
  lying 143-4
  naturalism as a set-up 149-50
  performatives and infelicities 146-7
  robots in the family 140-42
  and the study of human–robot
      interaction (HRI) 144-6
  technical glitch/error 151-2
Tinwell, A. 29-30, 131
Turing, A. 60
Turkle, S. xv 75, 103, 201, 208n

Uando (robot) 207
uncanny valley phenomenon 110, 111,
    123-4, 130-34, *132*
  historical reach 125-7
  and Buddhist-inspired thought 110,
    111
  and computer graphics 127-8
    capturing reality 129-30
    simulating reality 128-9
  and empathy with sociable robots 19-
    21, 35-6
    empathy and the uncanny valley
        29-35, 30n, 34n
    empathy *vs.* eeriness 21-5
    on the imaginative perception of
        emotion 25-9

understanding in human–robot interaction
    (HRI) 3-5, 15
  bodily movements and actions 10-14
  levels of interaction 5-9

Vallor, S. 58, 64n, 66
van Gelder, T. 60
virtue ethics 58, 61, 63-7, 64n, 73, 75, 92

Wallach, W. 60, 63
Wang, Y. 209, 210
Weber, J. 4, 208, 209, 210
Wegner, D.M. 33-4
Weizenbaum, J. 62n
Wettig, S. 166
Wheatley, T. 33
Wiener, N. 64n
Willson, M.A. 207
Wizard-of-Oz (WoZ) concept/method xx,
    144, 145-6
Wood, N. 109

Yamada, Y. 32-3
Yampolskiy, R.V. 190
Young, J.E. 209, 210

Zehendner, E. 166
Zemeckis, R. 125
Zen Buddhism, *see* Buddhist-inspired
    thought on human–robot
    interaction (HRI)